Pearson Australia
(a division of Pearson Australia Group Pty Ltd)
707 Collins Street, Melbourne, Victoria 3008
PO Box 23360, Melbourne, Victoria 8012
www.pearson.com.au

First published 2018 by Pearson Australia
2021 2020 2019 2018
10 9 8 7 6 5 4 3 2 1

Lead Publisher: Misal Belvedere
Project Manager: Michelle Thomas
Production Editors: Anji Bignell and Laura Pietrobon
Lead Development Editor: Fiona Cooke
Content Developer: Rebecca Wood
Development Editors: Haeyean Lee and Naomi Campanale
Editor: Fiona Maplestone
Designer: Anne Donald
Rights & Permissions Editors: Samantha Russell-Tulip and Jenny Jones
Senior Publishing Services Analyst: Rob Curulli
Proofreader: Trudi Ryan
Illustrator/s: DiacriTech
Printed and bound in Australia by Pegasus Media & Logistics

ISBN 978 1 4886 1931 1

Pearson Australia Group Pty Ltd ABN 40 004 245 943

Acknowledgements
The following abbreviations are used in this list: t = top, b = bottom, l = left, r = right, c = centre.

Cover image: Dennis Kunkel Microscopy/Science Photo Library
123RF: Ambient Ideas, p. 161(butterfly); jdeks, p. 116t; Duncan Noakes, p. 161(kookaburra); Frederik Johannes Thirion, p. 159(desert); Marco Tomasini, p. 161(wombat).
AAP: University of QLD/Ove Hoegh-Guldberg, p. 164b.
Alamy Stock Photo: Blickwinkel/McPHOTO/VLZ, p. 161(lyrebird); Bjorn Svensson/age fotostock, p. 163t; Universal Images Group North America LLC, p. 120; Genevieve Vallee, p. 159(river); Bosiljka Zutich, p. 163b.
Commonwealth of Australia: Based on Indicator: IW-38 Cane toad distribution. Department of Environment and Heritage 2005, p. 114.
Fotolia: Glen Gaffney, p. 161(parrot).
Getty Images: BIOPHOTO ASSOCIATES/Science Source, p. 44; Lester V. Bergman/Corbis Documentary, p. 16tr; Joe McDonald/Corbis Documentary, p. 116c.
imagefolk: FLPA/Terry Whittaker, p. 199t.
Pearson Education Ltd: Oxford Designers & Illustrators Ltd., p. 164t.
Yvonne Sanders: p. 151.
Science Photo Library: Dr Jeremy Burgess, p. 16br; Thomas Deerinck, NCMIR, pp. 49–50; Dennis Kunkel Microscopy, pp. i, xxiv, 16bl; Steve Gschmeissner, p. 16tl; ISM, p. 16cl; Louise Murray, p. 197; Dr Gopal Murti, pp. 1-2; Javier Torrent, VW PICS, p. 98–9.
Shutterstock: 580903, pp. 159(forest), 161(forest); Kristian Bell, p. 161(snake); bstoltz, p. 161(butterfly); Andrew Burgess, p. 103l; David Dirga, p. 188; Pichugin Dmitry, p. 159(mountain); Philip Ellard, p. 161(owl); Gallinago_media, p. 161(fox); Kjersti Joergensen, p. 199b; Piotr Krzeslak, p. 161(frog); Janelle Lugge, p. 103r; John McQueen, p. 116b; Nata-Lia, p. 161(mushroom); pisaphotography, pp. 48–9; SAP IBRAHIM, p. 199c; Eugene Sergeev, p. 161(glider); Joanne Weston, p. 165.

Contents

Contents

Module 3: Biological diversity

Module 4: Ecosystem dynamics

How to use this book

The *Pearson Biology 11 New South Wales Skills and Assessment Book* takes an intuitive, self-paced approach to science education that ensures every student has opportunities to practise, apply and extend their learning through a range of supportive and challenging activities. While offering opportunities for reinforcement of key concepts, knowledge and skills, these activities enable flexibility in the approach to teaching and learning.

Explicit scaffolding makes learning objectives clear, and there are regular opportunities for student reflection and self-evaluation at the end of individual activities throughout the book. Students are also guided in self-reflection at the end of each module. There are rich opportunities to take the content further with the explicit coverage of Working scientifically skills and key knowledge in the depth studies.

This resource has been written to the new New South Wales Biology Stage 6 Syllabus and addresses the first four modules of the syllabus. Each module consists of five main sections:

- key knowledge
- worksheets
- practical activities
- depth study
- module review questions.

Explore how to use this book below.

MODULE 1 Cells as the basis of life

Module opener

Each book is divided to follow the four modules of the syllabus, with the module opener linking the module content to the syllabus.

Biology toolkit

The Biology toolkit supports development of the skills and techniques needed to undertake practical investigations, secondary-sourced investigations and depth studies, and covers examination techniques and study skills. It also includes checklists, models, exemplars and scaffolded steps. The toolkit can serve as a reference tool, to be consulted as needed.

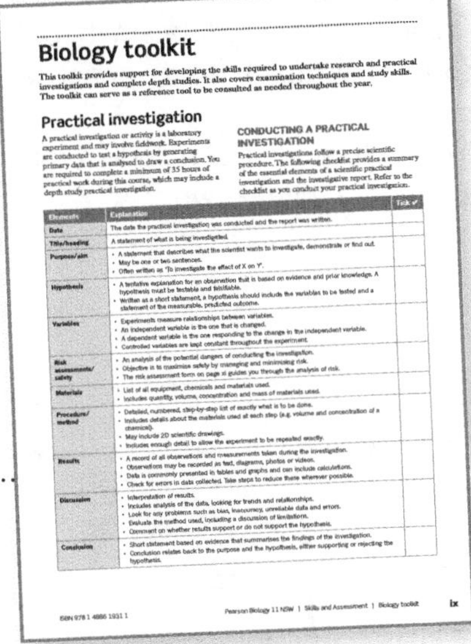
Biology toolkit

This toolkit provides support for developing the skills required to undertake research and practical investigations and complete depth studies. It also covers examination techniques and study skills. The toolkit can serve as a reference tool to be consulted as needed throughout the year.

Practical investigation

CONDUCTING A PRACTICAL INVESTIGATION

Key knowledge

Each module begins with a key knowledge section. This consists of a set of succinct summary notes that cover the key knowledge set out in each module of the syllabus. This section is highly illustrative and written in a straightforward style to assist students of all reading abilities. Key terms are in bold for ease of navigation. It also serves as a ready reference for completing the worksheets and practical activities.

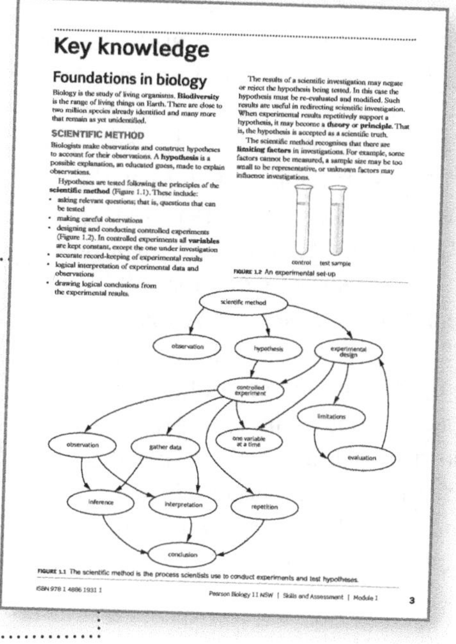
Key knowledge

Foundations in biology

SCIENTIFIC METHOD

FIGURE 1.1 The scientific method is the process scientists use to conduct experiments and test hypotheses.

Worksheets

A diverse offering of instructive and self-contained worksheets is included in each module. Common to all modules are the initial 'Knowledge review' worksheet to activate prior knowledge, a 'Literacy review' worksheet to explicitly build understanding and application of scientific terminology, and finally a 'Thinking about my learning' worksheet, which students can use for reflection and self-assessment. Other worksheet types provide opportunities to revise, consolidate and further student understanding.

All worksheets function as formative assessment and are clearly aligned to the syllabus. A range of questions building from foundation to challenging are included within the worksheets.

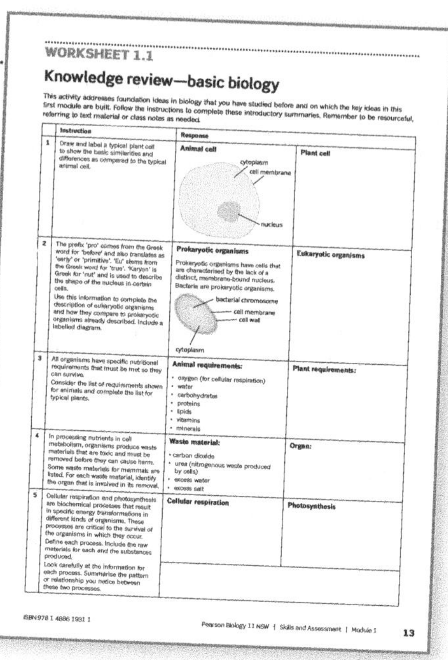
WORKSHEET 1.1

Knowledge review—basic biology

PRACTICAL ACTIVITY 1.1

Distinguishing cells—an observational activity

Practical activities

Practical activities give students the opportunity to complete practical work related to the various themes covered in the syllabus. All practical activities referenced in outcomes within the syllabus have been covered. Across the suite of practical activities, students have opportunities to design, conduct, evaluate, gather and analyse data, appropriately record results and prepare evidence-based conclusions. This can be done directly into the scaffolded practical activities. Students also have opportunities to evaluate safety and risk, and identify any potential hazards.

Each practical activity includes a suggested duration. Along with the depth studies, the practical activities meet the 35 hours of practical work mandated for Year 11 in the syllabus. Where there is key knowledge that will support the completion of a practical activity, students are referred back to it.

Like the worksheets, the practical activities include a range of questions, building from foundation to challenging.

Depth study

Each module contains at least one suggested depth study. The depth studies allow further development of one or more concepts found within or inspired by the syllabus. They allow students to acquire a depth of understanding and take responsibility for their own learning, and promote differentiation and engagement.

Each depth study allows for the demonstration of a range of Working scientifically skills, with all depth studies assessing the Working scientifically outcomes of Questioning and predicting, and Communicating. A minimum of two additional Working scientifically skills and at least one Knowledge and Understanding outcome are also assessed.

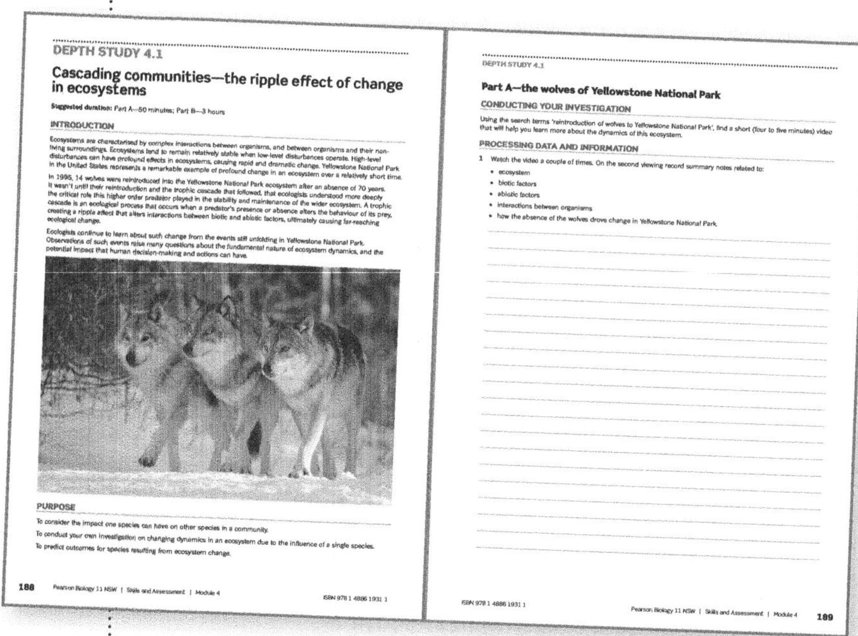
DEPTH STUDY 4.1

Cascading communities—the ripple effect of change in ecosystems

Part A—the wolves of Yellowstone National Park

 ISBN 978 1 4886 1931 1

Module review questions

Each module finishes with a comprehensive set of questions, consisting of multiple choice, short answer and extended response, which helps students to draw together their knowledge and understanding and apply it to these styles of question.

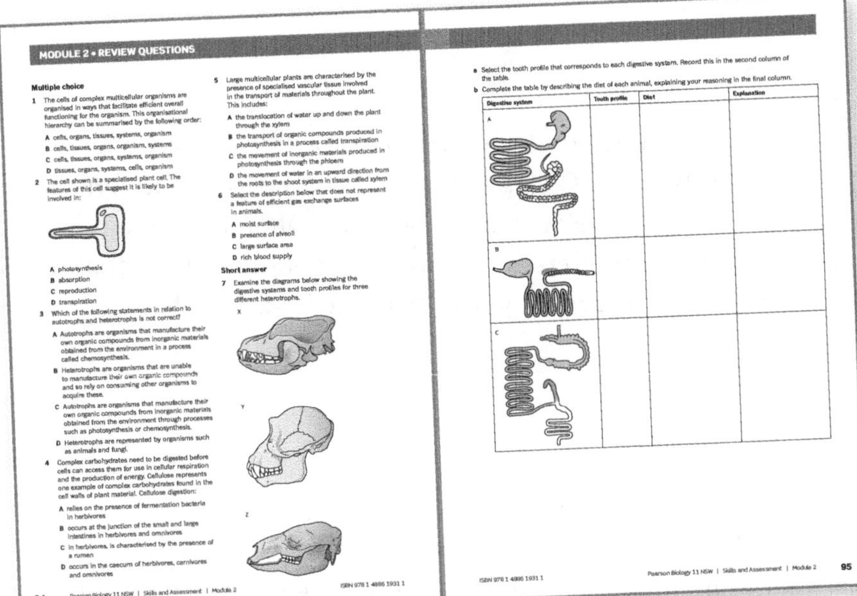
MODULE 2 • REVIEW QUESTIONS

Multiple choice

1 The cells of complex multicellular organisms are organised in ways that facilitate efficient overall functioning for the organism. This organisational hierarchy can be summarised by the following order:
A cells, organs, tissues, systems, organism
B cells, tissues, organs, organism, systems
C cells, tissues, organs, systems, organism
D tissues, organs, systems, cells, organism

2 The cell shown is a specialised plant cell. The features of this cell suggest it is likely to be involved in:
A photosynthesis
B absorption
C reproduction
D transpiration

3 Which of the following statements in relation to autotrophs and heterotrophs is not correct?
A Autotrophs are organisms that manufacture their own organic compounds from inorganic materials obtained from the environment in a process called chemosynthesis.
B Heterotrophs are organisms that are unable to manufacture their own organic compounds and so rely on consuming other organisms to acquire these.
C Autotrophs are organisms that manufacture their own organic compounds from inorganic materials obtained from the environment through processes such as photosynthesis or chemosynthesis.
D Heterotrophs are represented by organisms such as animals and fungi.

4 Complex carbohydrates need to be digested before cells can access them for use in cellular respiration and the production of energy. Cellulose represents one example of complex carbohydrates found in the cell walls of plant material. Cellulose digestion:
A relies on the presence of fermentation bacteria in herbivores
B occurs at the junction of the small and large intestines in herbivores and omnivores
C in herbivores, is characterised by the presence of a rumen
D occurs in the caecum of herbivores, carnivores and omnivores

5 Large multicellular plants are characterised by the presence of specialised vascular tissue involved in the transport of materials throughout the plant. This includes:
A the translocation of water up and down the plant through the xylem
B the transport of organic compounds produced in photosynthesis in a process called transpiration
C the movement of inorganic materials produced in photosynthesis through the phloem
D the movement of water in an upward direction from the roots to the shoot system in tissue called xylem

6 Select the description below that does not represent a feature of efficient gas exchange surfaces in animals.
A moist surface
B presence of alveoli
C large surface area
D rich blood supply

Short answer

7 Examine the diagrams below showing the digestive systems and tooth profiles for three different heterotrophs.

X

Y

Z

94 Pearson Biology 11 NSW | Skills and Assessment | Module 2

ISBN 978 1 4886 1931 1

a Select the tooth profile that corresponds to each digestive system. Record this in the second column of the table.
b Complete the table by describing the diet of each animal, explaining your reasoning in the final column.

Digestive system	Tooth profile	Diet	Explanation
A			
B			
C			

ISBN 978 1 4886 1931 1

Pearson Biology 11 NSW | Skills and Assessment | Module 2 95

Rating my learning

Rating my learning is an innovative tool that appears at the bottom of the final page of most worksheets and all practical activities. It provides students with the opportunity for self-reflection and self-assessment. It encourages them to look ahead to how they can continue to improve, and it helps them to identify focus areas for further skill and knowledge development.

The teacher may choose to use student responses to the 'Rating my learning' feature as a formative assessment tool. At a glance, teachers can assess which topics and which students need intervention for improvement.

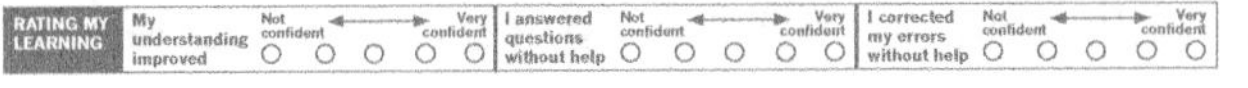

Icons and features

The 2018 New South Wales Biology Stage 6 Syllabus Learning Across the Curriculum content is addressed and identified:

GO TO ➤ **GoTo icons** are used to make important links to relevant content within the book.

The **safety icon** highlights significant hazards, indicating caution is needed.

The **safety glasses icon** highlights that protective eyewear is to be worn during the practical activity.

Highlight boxes focus students' attention on important information such as key definitions, formulae and summary points.

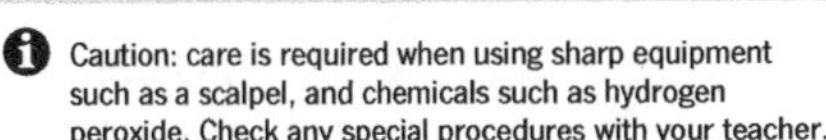

Teacher Support

Comprehensive answers and fully worked solutions for all worksheets, practical activities, depth studies and module review questions are provided via the *Pearson Biology 11 New South Wales* Teacher Support. An editable suggested assessment rubric for depth studies is also provided.

Pearson Biology 11 New South Wales

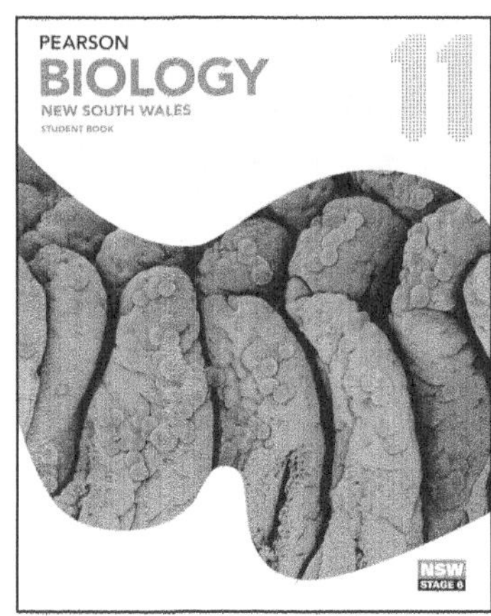

Student Book

Pearson Biology 11 New South Wales has been written to fully align with the 2018 New South Wales Biology Stage 6 Syllabus. The Student Book includes the very latest developments and applications of biology and incorporates best-practice literacy and instructional design to ensure the content and concepts are fully accessible to all students.

Skills and Assessment Book

The *Skills and Assessment Book* gives students the edge in preparing for all forms of assessment. Key features include a toolkit, key knowledge summaries, worksheets, practical activities, suggested depth studies and module review questions. It provides guidance, assessment practice and opportunities to develop key skills.

Reader+ the next generation eBook

Pearson Reader+ lets you use your *Student Book* online or offline on any device. Pearson Reader+ retains the look and integrity of the printed book. Practical activities, interactives and videos are available on Pearson Reader+, along with fully worked solutions for the Student Book questions.

Teacher Support

The Teacher Support available includes syllabus grids and a scope and sequence plan to support teachers with programming. It also includes fully worked solutions and answers to all *Student Book* and *Skills and Assessment Book* questions, including all worksheets, practical activities, depth studies and module review questions. Teacher notes, safety notes, risk assessments and laboratory technician's checklists and recipes are available for all practical activities. Depth studies are supported with suggested assessment rubrics and exemplar answers.

Pearson Digital

Access your digital resources at **pearsonplaces.com.au**
Browse and buy at **pearson.com.au**

ISBN 978 1 4886 1931 1

Biology toolkit

This toolkit provides support for developing the skills required to undertake research and practical investigations and complete depth studies. It also covers examination techniques and study skills. The toolkit can serve as a reference tool to be consulted as needed throughout the year.

Practical investigation

A practical investigation or activity is a laboratory experiment and may involve fieldwork. Experiments are conducted to test a hypothesis by generating primary data that is analysed to draw a conclusion. You are required to complete a minimum of 35 hours of practical work during this course, which may include a depth study practical investigation.

CONDUCTING A PRACTICAL INVESTIGATION

Practical investigations follow a precise scientific procedure. The following checklist provides a summary of the essential elements of a scientific practical investigation and the investigative report. Refer to the checklist as you conduct your practical investigation.

Elements	Explanation	Tick ✔
Date	The date the practical investigation was conducted and the report was written.	
Title/heading	A statement of what is being investigated.	
Purpose/aim	• A statement that describes what the scientist wants to investigate, demonstrate or find out. • May be one or two sentences. • Often written as 'To investigate the effect of X on Y'.	
Hypothesis	• A tentative explanation for an observation that is based on evidence and prior knowledge. A hypothesis must be testable and falsifiable. • Written as a short statement, a hypothesis should include the variables to be tested and a statement of the measurable, predicted outcome.	
Variables	• Experiments measure relationships between variables. • An independent variable is the one that is changed. • A dependent variable is the one responding to the change in the independent variable. • Controlled variables are kept constant throughout the experiment.	
Risk assessments/ safety	• An analysis of the potential dangers of conducting the investigation. • Objective is to maximise safety by managing and minimising risk. • The risk assessment form on page xi guides you through the analysis of risk.	
Materials	• List of all equipment, chemicals and materials used. • Includes quantity, volume, concentration and mass of materials used.	
Procedure/ method	• Detailed, numbered, step-by-step list of exactly what is to be done. • Includes details about the materials used at each step (e.g. volume and concentration of a chemical). • May include 2D scientific drawings. • Includes enough detail to allow the experiment to be repeated exactly.	
Results	• A record of all observations and measurements taken during the investigation. • Observations may be recorded as text, diagrams, photos or videos. • Data is commonly presented in tables and graphs and can include calculations. • Check for errors in data collected. Take steps to reduce these wherever possible.	
Discussion	• Interpretation of results. • Includes analysis of the data, looking for trends and relationships. • Look for any problems such as bias, inaccuracy, unreliable data and errors. • Evaluate the procedure used, including a discussion of limitations. • Comment on whether results support or do not support the hypothesis.	
Conclusion	• Short statement based on evidence that summarises the findings of the investigation. • Conclusion relates back to the purpose and the hypothesis, either supporting or rejecting the hypothesis.	

RISK ASSESSMENT FORM

Five levels of safety should be considered in an investigation. The inverted pyramid ranks these levels in order of importance. The school and your teacher are responsible for reducing most of these risks. You as a student scientist can take measures to reduce the risks shown at the bottom of the hierarchy.

Complete the risk assessment form to identify possible risks for which you can take responsibility, and to think of ways you can reduce risks to create a safe environment.

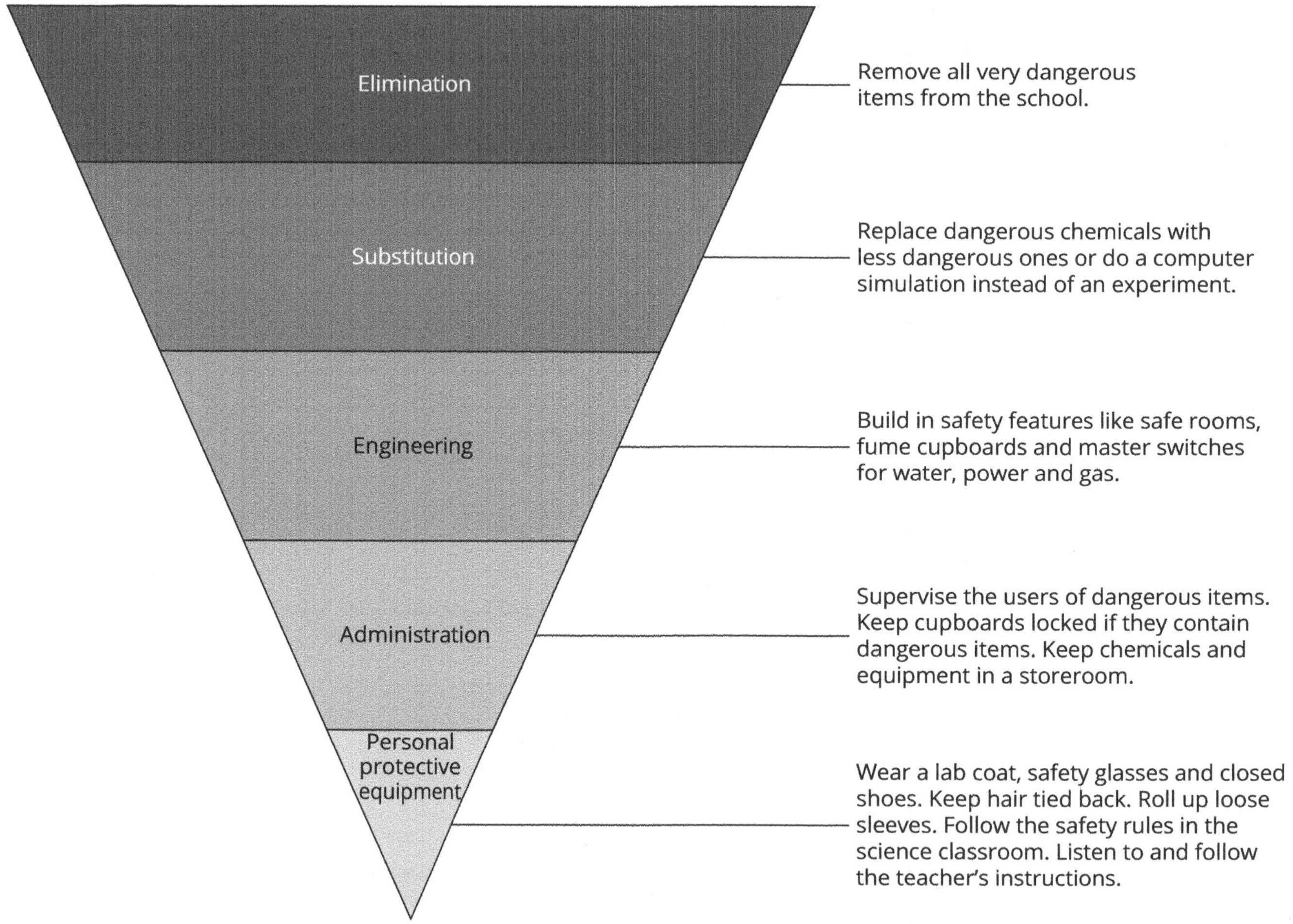

ISBN 978 1 4886 1931 1

Risk assessment form		
What activity are you doing?		
Title or description of the investigation:		
List ...	**Identify any risks**	**State how you will ...**
equipment you will be using:		use each piece:
chemicals you will be using:		carefully use each chemical: carefully dispose of the chemicals:
ethical issues you need to consider:		ethically use animals in the laboratory: ethically use human participants in the investigation:
outdoor or fieldwork activities:		reduce these risks:
any other possible risks:		reduce these risks:

EXAMPLES OF PRACTICAL REPORTS

It can be difficult to gauge whether you have attained a high standard in your completed practical activity report. Looking at sample practical reports can help you identify what is required. Two sample practical reports are provided: one is prepared to a high standard, while the second has room for improvement. The annotations draw your attention to key points to note on each practical report. These points are also reflected in the checklist so you are able to use this as a tool to evaluate whether all requirements of the practical have been met.

High standard practical report

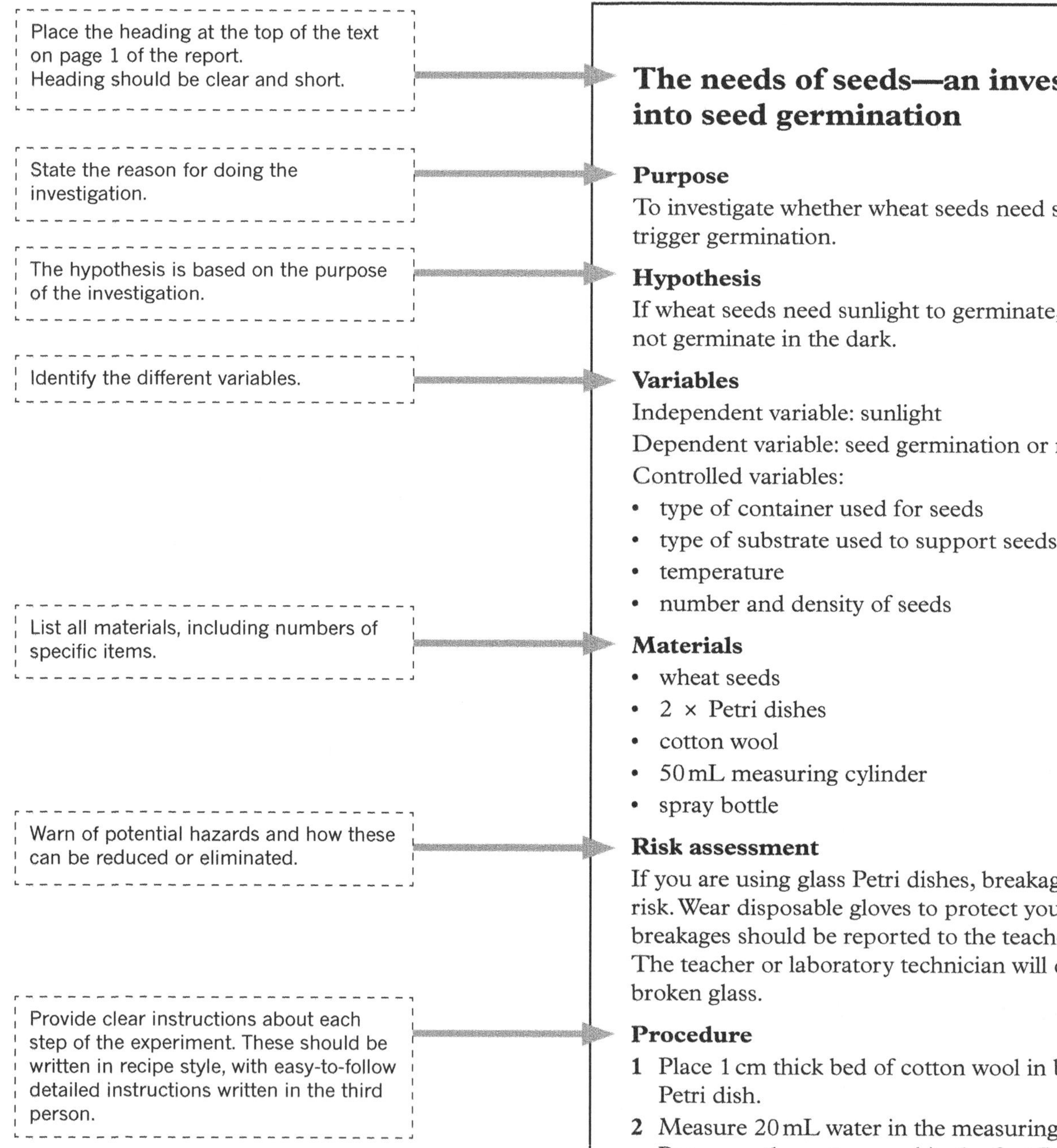

The needs of seeds—an investigation into seed germination

Purpose

To investigate whether wheat seeds need sunlight to trigger germination.

Hypothesis

If wheat seeds need sunlight to germinate, then they will not germinate in the dark.

Variables

Independent variable: sunlight

Dependent variable: seed germination or not

Controlled variables:

- type of container used for seeds
- type of substrate used to support seeds
- temperature
- number and density of seeds

Materials

- wheat seeds
- 2 × Petri dishes
- cotton wool
- 50 mL measuring cylinder
- spray bottle

Risk assessment

If you are using glass Petri dishes, breakage is a potential risk. Wear disposable gloves to protect your hands. All breakages should be reported to the teacher immediately. The teacher or laboratory technician will dispose of broken glass.

Procedure

1. Place 1 cm thick bed of cotton wool in base of each Petri dish.
2. Measure 20 mL water in the measuring cylinder. Pour over the cotton wool in the first Petri dish until uniformly wet. Repeat this for the second Petri dish.
3. Spread 20 wheat seeds evenly on the cotton wool bed of each Petri dish.
4. Place one Petri dish on the windowsill where it will receive sunlight. Place the second in a dark cupboard nearby.
5. Leave the Petri dishes for seven days. Use the water spray bottle to spray the two Petri dishes every second day to ensure they do not dry out.
6. Compare the seeds of the two set-ups after seven days.

 ISBN 978 1 4886 1931 1

Results

Wheat seeds placed in light and dark conditions did not show a difference in the mean number of seeds germinated over a period of seven days (Table 1).

TABLE 1 Number of wheat seeds germinated in light and dark conditions over seven days

Group	Number of seeds germinated	
	Light	Dark
1	18	17
2	17	19
3	18	16
4	17	17
5	12	14
Mean	16.4	16.6

- Record your results in an appropriate format. Tables are useful for recording experimental data. Graphs, diagrams and photographs are also useful approaches for recording data.

- If you take several readings or a reading from several groups, calculating the mean can be useful to see overall patterns.

Discussion

The data from each set of experiments follows a similar pattern. The majority of wheat seeds germinated, regardless of exposure to light or dark.

The light and dark treatments for Group 5 both have a lower number of germinating seeds than the other groups. As the Group 5 seeds came from the same source as the other groups' seeds, variables other than the seeds need to be considered when accounting for the difference in germination rates between groups. There are several possibilities that might account for the lower rate of germination. For example, the seeds may not have received adequate water. Another reason could be the handling of the seeds by contaminated hands. Such variables should be controlled in future experiments by ensuring all seeds receive equal quantities of water and by wearing disposable gloves when handling the seeds.

The discussion focuses on the interpretation of the experimental data. What do the results show? Are there any unexpected results? How can we account for this? Evaluation of the procedure is also included here.

Conclusion

This investigation demonstrated that wheat seeds do not require sunlight in order to germinate. The hypothesis is not supported. The experimental evidence shows that in every instance, wheat seeds germinated when placed in the dark for seven days.

The conclusion relates back to the purpose and states whether the hypothesis was supported or not supported. It also outlines the experimental evidence to support this.

Avoid terms such as 'always' and 'never' in scientific writing. Refer to what the data shows—all claims must be supported by the evidence.

Low standard practical report

- The title should be written as a statement, not a question.

Do seeds need sunlight to germinate?

Purpose

I plan to see if seeds need sunlight to germinate.

- The purpose is not specific enough and is not followed by a hypothesis.
- A list of variables is missing.
- Risk assessment notes are absent.

Materials

I'll need seeds, containers and water.

- Materials should be listed as dot points, with numbers of items and volumes of substances provided.
- Appropriate terminology should be used.

Procedure

Line each container with some cotton wool, then add some seeds and water.

- This procedure is too brief and non-specific. It must be a set of detailed step-by-step instructions that can be easily repeated by someone else. It should include precise quantities (e.g. numbers of seeds for each Petri dish).
- Diagrams can be a useful way of providing direction in the procedure.

Results

Some of the seeds germinated and others didn't. We didn't water the seeds regularly, so this might have had something to do with the results.

- These results are too brief and don't include any quantitative or qualitative data.
- Include a table or graph of results, quantitative or qualitative as appropriate (e.g. a table showing the number of seeds that germinated in the light and in the dark would be useful in this investigation).
- Scientific reports should be written in the third person, not the first person (which uses 'I').
- Keep the language in a scientific report formal.

Discussion

Some of the seeds germinated so they must have had all of the requirements for germination, which included light. Some of the seeds also germinated in the dark.

This discussion is too vague—it doesn't provide any detailed explanation to account for the results.

Conclusion

We succeeded in completing this investigation.

This conclusion is too vague—it doesn't respond specifically to the purpose of the investigation, nor state whether the hypothesis was supported or not supported. It doesn't use the experimental results to support the conclusion.

Secondary-sourced investigation

This section guides you in conducting a secondary-sourced investigation. For assistance with conducting a practical activity, see pages ix–xiv.

A major secondary-sourced investigation is also often known as a research project. Such investigations require you to think carefully about the topic, find, collect and organise information, analyse and synthesise findings and present your ideas. The investigation process is summarised in the following flowchart. An investigation is not necessarily a straightforward linear process as shown in the flowchart. You can move back and forth between steps as needed.

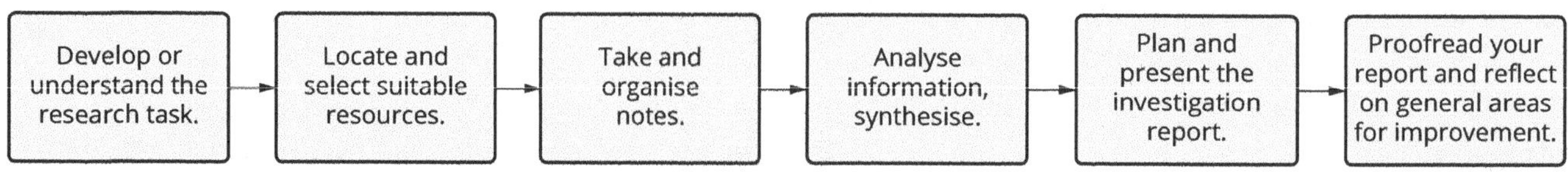

THE INVESTIGATION TASK

It is important to understand the breakdown of investigation questions or tasks. This helps you to write your own tasks and better understand the requirements of those written by others.

As you develop an investigation task, be aware of the depth of thinking it will require. The following chart provides support in writing tasks at different levels of thinking or complexity. It also provides some key words to help target tasks at these thinking levels and gives some examples of questions.

When developing your question or task, be conscious of the level at which you are pitching it. When you are writing your own questions for a depth study investigation, they should generally be at the analysis level.

 ISBN 978 1 4886 1931 1

Level of question complexity	Type of thinking	Words that may be used		Examples of questions
Simple	**Retrieval:** remembering, producing information on demand	• who • where • list • show • describe • select • complete • define	• what • when • label • demonstrate • name • state • recognise	**1** Define the term 'diffusion'. **2** What is the function of an enzyme? **3** List the structures in the mammalian circulatory system. **4** What are the products of photosynthesis?
	Comprehension: the ability to understand information	• who • where • explain • represent • show how	• what • when • summarise • draw • describe	**1** Represent the process of osmosis with a labelled diagram. **2** Why are the processes of photosynthesis and cellular respiration described as inverse reactions? **3** Explain how enzymes may be described using the 'lock-and-key' model. **4** What happens to enzymes when they are subjected to extremes of pH?
	Analysis: scrutinising and breaking something into its smaller parts, including: • comparing • classifying • identifying errors • concluding • predicting • judging	• why • categorise • contrast • sort • organise • generalise • evaluate • edit • assess • judge	• how • compare • distinguish • discriminate between • deduce • critique • diagnose • identify errors • identify misunderstandings	**1** Organise the following in order from simplest to most complex: organ, specialised cell, system, tissue. **2** Compare open and closed transport systems in animals. **3** Compare and contrast physical and chemical digestion. **4** Organise the structures of the mammalian cardiovascular system from those with high blood pressure to those with low blood pressure. Present this information in a diagram or graph.
Complex (requiring more thinking)	**Application:** using knowledge in new situations, including: • testing a hypothesis • solving a problem • experimenting and using data • decision making	• why • investigate • find out about • test • solve • develop • decide	• how • research • experiment • predict • adapt • judge	**1** Many scientists believe that limiting human population growth is necessary to control environmental damage. Construct an argument for or against this statement. **2** Investigate the effect of vitamin B12 deficiency in humans. **3** Research the link between rising global temperatures and coral bleaching on the Great Barrier Reef.

RESOURCES

The resources you refer to in the investigation may be primary and/or secondary.

Primary sources of information are from investigations that you have conducted yourself. Examples of primary sources are results from your experiments, reports of your scientific investigations, photographs you have taken, and specimens or artefacts that you have collected. Secondary sources of information are from investigations that have been conducted by others. Secondary sources include peer-reviewed articles, textbooks, biographies, documentaries, newspaper articles and websites.

Refer to the two checklists below. The first shows features to look for when assessing and selecting the best sources for your investigation. The second shows the information that is required for the references section of your report.

Selecting resources for the investigation	Tick ✔
The resource is:	
• **credible** and I can identify the author, author's expertise and publisher	
• **current** because the date of publication of the material is provided and is recent	
• **factual** and I know that it is objective material and not biased	
• **accurate** and all information is correct	
• **relevant** and covers the area I am investigating	
• **readable** and neither too simple nor too complex in its coverage of the material	

Examples of information required for references and bibliographies
Article in scientific magazine Author, initials. (year). Title of article. *Journal title, volume number*(issue number), page numbers. Pritchard, H. D. (2017). Asia's glaciers are a regionally important buffer against drought. *Nature, 545*(7653), 169–174.
Book Author, initials. (year). *Title of book* (edition, if not first). City: Publisher. Rickard, G. *et al.* (2017). *Pearson Science Student Book 9.* Melbourne: Pearson Education. Where there are more than two authors, list first author then write *'et al.'* meaning 'and others'.
Internet Author, initials/name of organisation. (year). *Title of webpage or web document.* Retrieved from <URL>, date accessed Royal Society of Biology. (2017). *Biology for beginners.* Retrieved from https://www.rsb.org.uk/, 12/10/17

Always remember to record the above details for each resource you use. It is important to accurately cite resources in your reference section and it can be time-consuming and difficult to find this information later.

NOTE-TAKING AND ORGANISING NOTES

Note-taking and organising requires skill. Good note-taking helps you avoid plagiarism and provides excellent information to support your scientific report writing. Plagiarism is when you take someone else's ideas and words and present them as your own work. You plagiarise if you copy sections or sentences from sources or you cut and paste from the Internet. It is acceptable to use the ideas of others but you must state clearly where the information has come from in the references.

The examples in the following table are of original text, plagiarised text and acceptably rephrased text.

Original text	Plagiarised text	Rephrased text
Dogs have sweat glands on their feet. Dogs pant when they are hot because their sweat glands are not sufficient to cool down their bodies. In addition, their tongues allow the water from their bodies to evaporate and cool down their bodies.	Dogs have their sweat glands on their feet. When dogs get hot they pant because their sweat glands are not enough to cool down their bodies. Their tongues let the water from their bodies evaporate and cool their bodies.	Dogs pant in order to cool down. Water evaporates from their tongues and this lowers their body temperature. They can't get cool enough through just the sweat glands on their feet.

There are various approaches to effective note-taking. Whatever technique you use, try to keep notes brief and focus on key points. Some examples include:

- dot point summary
- underlining or highlighting text
- labelled diagram
- flowchart—show sequences
- concept map—show connections between ideas
- Venn diagrams—show similarities and differences
- table—may incorporate any of the other note-taking techniques. Tables are useful for summarising longer and more complex information that has subparts. Adapt the table to suit your style and the task. The following sample table (partially completed) shows how this technique can be used to take notes for a secondary-sourced investigation.

 ISBN 978 1 4886 1931 1

Secondary-sourced investigation			
Many scientists believe that limiting human population growth is necessary to control environmental damage. Construct an argument for or against this statement.			
	Source 1 (e.g. book) Title: Author: Publisher's name: Publisher's location: Date of publication:	**Source 2** (e.g. Internet) Title: Author: URL: Date accessed:	**Source 3** (e.g. science journal) Author: Date: Title of article: Journal title: Volume number: Pages:
Population growth trends		**Human population growth** 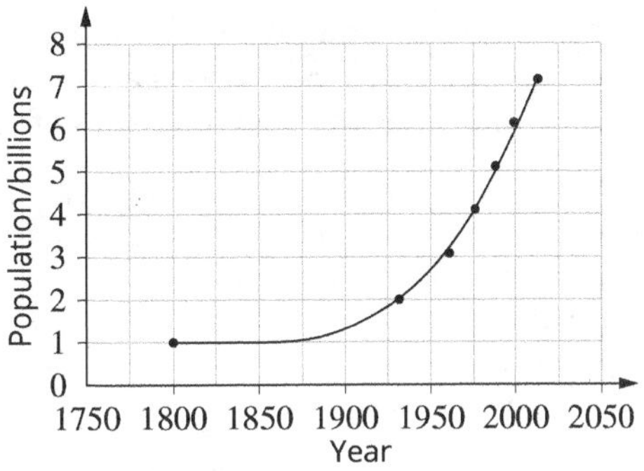 Population/billions (0–8) Year (1750 1800 1850 1900 1950 2000 2050)	
Impact of population growth on the environment	• growth of cities • demand for resources		
Reasons to control and limit population growth	• increased demand for resources for human survival is unsustainable and causing environmental damage		
Reasons not to limit population growth but to allow natural population growth			• other factors contributing to environment damage; land-use policies, e.g. poor land use

SCIENTIFIC WRITING

Scientists have a particular writing style. Your investigation should use this distinctive style to communicate your ideas. Scientific writing is:

- objective—describes events rather than what people think or feel and is as free as possible of bias or personal opinion
- precise—avoids exaggeration and uses qualified language
- formal—scholarly language rather than colloquial or everyday language
- concise—conveys information in short, clearly understandable sentences without unnecessary information
- simple—uses short sentences where possible
- predominantly written in passive voice, although sometimes active voice can be used to avoid any confusion
- structured to include headings, tables, diagrams and mathematical calculations.

Examples of unscientific and scientific writing are demonstrated in the following table.

Unscientific writing	Scientific writing
Subjective, biased writing: • The results were fantastic. • This produced a disgusting odour. • The breathtakingly beautiful bowerbird...	**Objective, unbiased writing:** • The results showed... • This produced a pungent odour... • The golden bowerbird...
Exaggerated writing: • The object weighed a huge amount. • The magnesium burst into huge flames. • Millions of ants swarmed over...	**Accurate, precise writing:** • The mass of the object was 250 kg. • The magnesium burnt vigorously. • Ants swarmed all over...

Everyday, informal language:	Formal language:
• The bacteria passed away. • The results don't... • We guessed that... • Previous researchers were slack and missed...	• The bacteria died. • The results do not... • It was hypothesised that... • Previous researchers have not found...
Active voice: • We recorded oxygen levels every hour. • We put 50 g of solute in a conical flask containing distilled water, and then we slowly added 1 mol L^{-1} hydrochloric acid.	**Passive voice:** • The oxygen level was recorded hourly. • 50 g of solute was placed in a conical flask containing distilled water, and then 1 mol L^{-1} of hydrochloric acid was slowly added.

PRESENTING THE INVESTIGATION

Scientific findings may be presented in a variety of ways. A common presentation format at science conferences is a poster. Posters can get ideas across to a large audience in an organised, concise and creative way. Other common presentation formats are essays, reports, oral presentations and articles. Each presentation format has its own conventions. The following table summarises the characteristics of a number of presentation formats.

Presentation format	Characteristics/inclusions	
Poster	• Balance of text and visuals • Title, subheadings • Balanced layout • Captions for figures and tables	• References • Hierarchy of font size according to subheading level • Consistent font style—no more than three fonts
Report/article	• Structured with introduction, paragraphs, conclusion • Includes subheadings	• Mainly text • Can include diagrams, graphs, tables
Essay	• Structured with an introduction, paragraphs, conclusion • Introduction states focus of essay • Each paragraph makes a new point supported by evidence	• Each paragraph links back to last paragraph • A text-style presentation format—visuals at end in appendix • Conclusion draws all ideas together but does not include any new information
Oral presentation	• Needs to be engaging • Refer to cue cards but do not read from them • Watch audience as you speak	• Stand still and don't fidget • Look at audience and appear confident

PROOFREADING

After you have completed the investigation and prepared your presentation, it is important to think about and check what you have done.

Proofread your work to minimise errors and maximise communication of the ideas from your investigation. Use the following questions as a proofreading checklist.

Proofreading checklist	Tick ✓
Have I investigated the question fully?	
Have I expressed myself clearly to communicate my ideas well?	
Have I used the scientific writing style?	
Have I included data analysis?	
Have I checked spelling, punctuation and grammar?	
Have I included references?	
Have I met the requirements of the presentation format?	

ISBN 978 1 4886 1931 1

Depth study

A depth study is an investigation that allows you to look in more detail at a particular area of interest in the syllabus. The depth study gives you the chance to gain greater understanding of concepts and is designed so you take more responsibility for your own learning.

It is expected that you demonstrate the use of a variety of scientific skills. Depth studies vary and may include practical work, fieldwork reports, research assignments, and may be based on primary or secondary data or sources. To fulfil the minimum 15-hour time requirement for depth study work, your teacher may ask you to complete one very detailed depth study or a number of smaller depth studies.

PLANNING THE DEPTH STUDY

Careful planning before actually starting the investigation will help you to be clear about what you will do and how you will do it. Think of yourself as taking on the role of your teacher or a scientist—identify the depth study investigation topic, decide how it will be investigated and presented, manage time through the investigation and check that all syllabus requirements are met.

Clarify your ideas for your depth study by following these steps.

1 List all areas that interest you for your depth study. For example, extending the work you've already completed on osmosis by considering how this process works to maintain cytoplasm concentrations in freshwater and saltwater fish or further investigating enzymes by considering their role in yoghurt production or probiotic activity.	**My areas of interest are:** __________ __________ __________ __________
2 Do some quick research into each area of interest. Identify which of these gives you scope for development of a depth study.	**My chosen area of investigation is:** __________ __________
3 Decide on the procedure you will use to conduct the depth study: • a practical investigation - may begin with fieldwork - may be a secondary-sourced investigation - may include data analysis • a secondary-sourced investigation - may be expository (explain something) - may be an argument based on evidence - may include data analysis - may involve an investigation of emerging technologies • creating - may design and construct a working model - may create a portfolio	**My procedure will be:** __________ __________ __________ __________ __________ __________ __________
4 Decide how you will present the depth study. Select a presentation format that suits your type of investigation: • a practical investigation (practical report) • a secondary-sourced investigation - documentary - media report - literature review - visual presentation - journal article - essay - environmental management plan • designing, inventing or creating - portfolio - working model	**I will present my depth study by:** __________ __________ __________ __________ __________ __________ __________

5 Now that you have thought about what you will investigate and how you will conduct the depth study, think about how you can best use your time. It is easy to spend most of the time conducting the practical or finding resources for the secondary-sourced investigation. This may leave little time to analyse data and prepare your presentation. Refer to the suggested planning timeline. Plan the depth study — 10% Conduct the investigation — 40% Organise information and analyse data — 30% Present information and proofread work — 20% Time allocation	**My depth study begins on** __________ **and is to be completed by** __________ Fill in the timeline to plan how you will allocate your time to complete the depth study, including specific dates.
6 Assessment requirements Depth study must: • address the Working scientifically skills 'Questioning and predicting', and 'Communicating' • address a minimum of two additional Working scientifically skills: - Planning investigations - Conducting investigations - Processing data and information - Analysing data and information - Problem solving • include at least one Knowledge and understanding outcome	**My depth study will include these Working scientifically skills:** **1** Questioning and predicting **2** Communicating **3** __________ **4** __________ **My depth study covers this/these Knowledge and Understanding outcomes:** __________ __________

Study skills

There are a variety of techniques and strategies to help you study. You may find that you use different strategies in different situations. For example, you may prefer to highlight key phrases in your notebook throughout the year but make summaries of topics before an examination. The strategies you choose will depend on personal preference and may not be the same as those used by classmates.

Effective study skills involve more than the learning strategies you use. Equally important is when you use those skills. It is more effective to apply study skills throughout the year, revising and consolidating your knowledge as you progress through the course, rather than doing a rushed cram just before the examination. Revise your work regularly. Being organised and setting up a study plan is key to reducing your stress.

GETTING ORGANISED

To get yourself organised, try the following steps:

- Use a diary to write down all homework and assessment tasks as soon as you get them. Note due dates and what is required.
- Be specific about the tasks you need to do. Rather than writing 'do biology', it is more effective to note things such as which questions to answer and which page to look at in your student book.
- Write a list of everything you need to do each day. Tick off or cross out items as you complete them.
- Break down larger tasks into smaller separate parts that are manageable.
- Make sure your lists and planners are realistic. Do not set yourself more than you can actually do.

STUDY TECHNIQUES

Studying requires concentration. Remove any distractions and factor in some breaks. Allow a 10-minute break every hour. Vary your study technique depending on the content to be learned and your personal preference. Although you may have already found a study technique that works for you, also consider the following options.

ISBN 978 1 4886 1931 1

Study technique	Tips
Highlighting FUNGI Fungi often look like plants but do not use photosynthesis. Instead they feed on dead and decaying material, breaking it down further and helping chemical elements to return to the natural environment. Mushrooms, toadstools, yeasts and moulds are different types of fungi.	• Highlight or underline key points as you read your notes or text.
Summary notes *ENZYMES* *Enzymes are:* • *composed of protein* • *substrate-specific* • *denatured by exposure to excessive heat* • *denatured by exposure to extremes of pH.*	• Create a list of key headings and add some dot points about each heading. • Write your own summary of the key ideas in each chapter. • Use headings and subheadings. • Underline key words and key phrases. • Use simple diagrams. • The most effective chapter summaries are clear, concise and uncluttered.
Diagrams 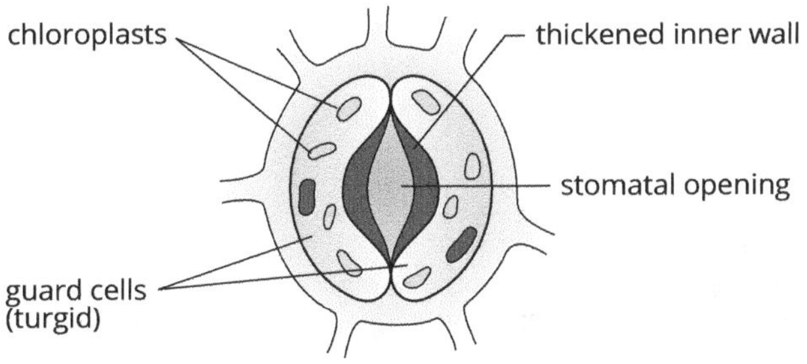 chloroplasts thickened inner wall stomatal opening guard cells (turgid)	• Diagrams can be used as a summary of key concepts. • They are useful memory triggers. • Diagrams cover a lot of information in a visual way, with minimal text.
Concept maps 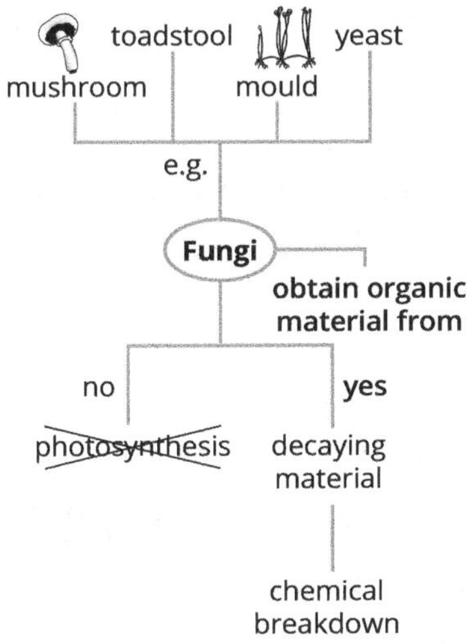 toadstool yeast mushroom mould e.g. **Fungi** **obtain organic material from** no **yes** photosynthesis decaying material chemical breakdown	• Concept maps are a great way of connecting key terms and ideas in a simple and ordered way. • Use the lines that connect ideas to note the relationships between the words and phrases. • They may include text and images. • They may be a simple or more complex summary tool. • Concept maps and other graphic organisers such as Venn diagrams and flowcharts show how information is connected and help deepen understanding.
Tables	• Tables are useful to show relationships between different factors. • Information is uncluttered.
Mnemonic devices *King Penguins Can Only Fly Going South* *... triggers your memory for ...* *Kingdom, Phylum, Class, Order, Family, Genus, Species*	• Mnemonic tools help you remember information. • To remember the classification of living things, memorise the sentence shown, which contains the initial letters for the classification types.
Glossary *Cell—structural unit of living things.*	• Compile your own glossary—writing the terms and definitions will make them easier to remember. • Add images to help you remember. • Write each term in a sentence. • Memorise these terms and definitions.

Table 1.2 Weed population

Weed Type	*Sunny lawn*	*Shaded lawn*
A	*20 plants*	*7 plants*
B	*9 plants*	*12 plants*

Trigger words *Enzyme engaging with substrate—two theories* • *Lock-and-key model* • *Induced fit model*	• Write down the key words associated with a topic or theme. • Trigger words are useful in helping to remember other related words and ideas.
Repeating information aloud	• This is a good way to remember and helps you to slow down and absorb the information.
Practice	• Do as many review questions and old exams as possible.
Flash cards Osmosis: Diffusion: process in which particles move from an area of high concentration to an area of low concentration; passive	• Making flash cards helps your understanding. • Make your own cards. • Write a question on one side and the answer on the back. • Cards can include definitions, brief explanations, diagrams, equations and graphs.
Teaching someone	• Teach friends or family members. • Teaching a difficult concept to someone means you must first understand the concept yourself.
Handwriting notes	• Hand-write rather than type summary notes. • Remember, the examination requires you to write answers. • Practise writing for long stretches of time and make sure your writing is legible.
Responding to feedback and self-correcting	• Check through all feedback from your teacher. • Highlight what was right or wrong. • Attempt to identify where you have errors and rework the answer to get it right.

EXAMINATION PREPARATION

In the weeks before the examination, begin your exam preparation. The earlier you begin revising, the easier it will be. It is also helpful to begin practising exam-style questions as early as possible, not just in the weeks ahead of the exam.

Like most skills, practice will improve your ability to do exams and to handle different types of exam questions. Doing practice exams is vital because you gain experience in:

- using reading time effectively before starting to write
- allocating the right amount of time to each question
- working to a time limit
- reading and interpreting questions
- understanding what is required for each question
- planning answers
- deciding on relevant information
- proofreading/checking over your own answers
- writing efficiently for the duration of the exam.

Use the following checklist as a reminder of your study program.

Study program checklist	Tick ✔
Have I revised all areas of the course?	
Have I highlighted important points?	
Have I made a summary of the important points in each topic?	
Have I read over my revision notes?	
Have I looked at and worked through sample exam papers?	
Have I answered practice questions in the appropriate time limit?	

ISBN 978 1 4886 1931 1

EXAMINATION STRATEGIES

Familiarise yourself with the conditions of the examination well before the day you sit the exam. You should know:

- the number of exams for the subject
- the amount of reading time allowed in the exam before writing begins
- the amount of writing time allocated
- any particular equipment allowed and/or required, such as a calculator, pencils, pens and ruler
- strategies to tackle the exam. Exam strategies are listed in the following table.

Exam strategies
Reading time • Remember that no writing at all is allowed during this time—no note-taking, no highlighting, no underlining. • Read the instructions. • Read through the short answer questions first. • Read the multiple choice questions next. • Read the remaining questions.
Writing time • Begin with the multiple choice questions. • Answer every multiple choice question, even if you can only make an educated guess. • If you are unsure of an answer to a multiple choice question, mark it so you can come back to it if time allows. • Attempt the short answer questions next. • Attempt the easiest short answer questions first and work your way to the more challenging questions. • Attempt all other questions next.
Tips for answering questions • Carefully read each question, underlining key words. • Be aware that most questions are structured so they become more challenging towards the end. You may not be able to answer the last part of a question but you will have earned most of the marks by answering earlier parts. • Look carefully at any diagrams, pictures, tables and graphs and make sure you understand their relevance to the questions involved. • For questions with graphs, read the graph title and the labels on the axes carefully, so that you can establish the relationship the graph is showing. • For questions with tables, read the headings on the columns and rows carefully, so that you can analyse the content of the table effectively. • Check for the key words in a question. Highlight them but don't colour the whole question. • Plan your answers before you write, remembering to address the exam criteria. • For questions with parts, read the whole question first. This gives you an overall picture of the question. It will also help to ensure that you do not repeat yourself in subsequent parts of the question. • Make sure you actually answer the question that is asked. • Once you have answered the question, re-read your answer and then re-read the question, to ensure that you have actually answered all of the question. • When writing a definition, don't use the word you are defining in your definition. • If giving values from a graph, use a ruler to line up points with the axes so you can be accurate, and always include units in your answer. • Be sure to attempt all questions. • Read over your answers to pick up careless errors—the mind is faster than the hand, and you may not always write what you intend (especially when you have limited time). • Write legibly. Exams are scanned and marked online. If the assessor can't read your answer, they can't mark it as correct. • Keep an eye on the time. • Never leave an exam early. Use any spare time to re-read and check your answers.
Exam cues • The number of marks allocated to a question provides a clue about how much you are expected to write. Two marks usually means you need to make a minimum of two points. • The number of lines allowed for the answer indicates the length of the expected answer. If your writing is large, you may need to turn the page over and continue on the back. Make sure you indicate that the examiner must turn to the back of the page for the rest of the answer. Where the answer continues, clearly state that it is the continuation of the question and state the question number.

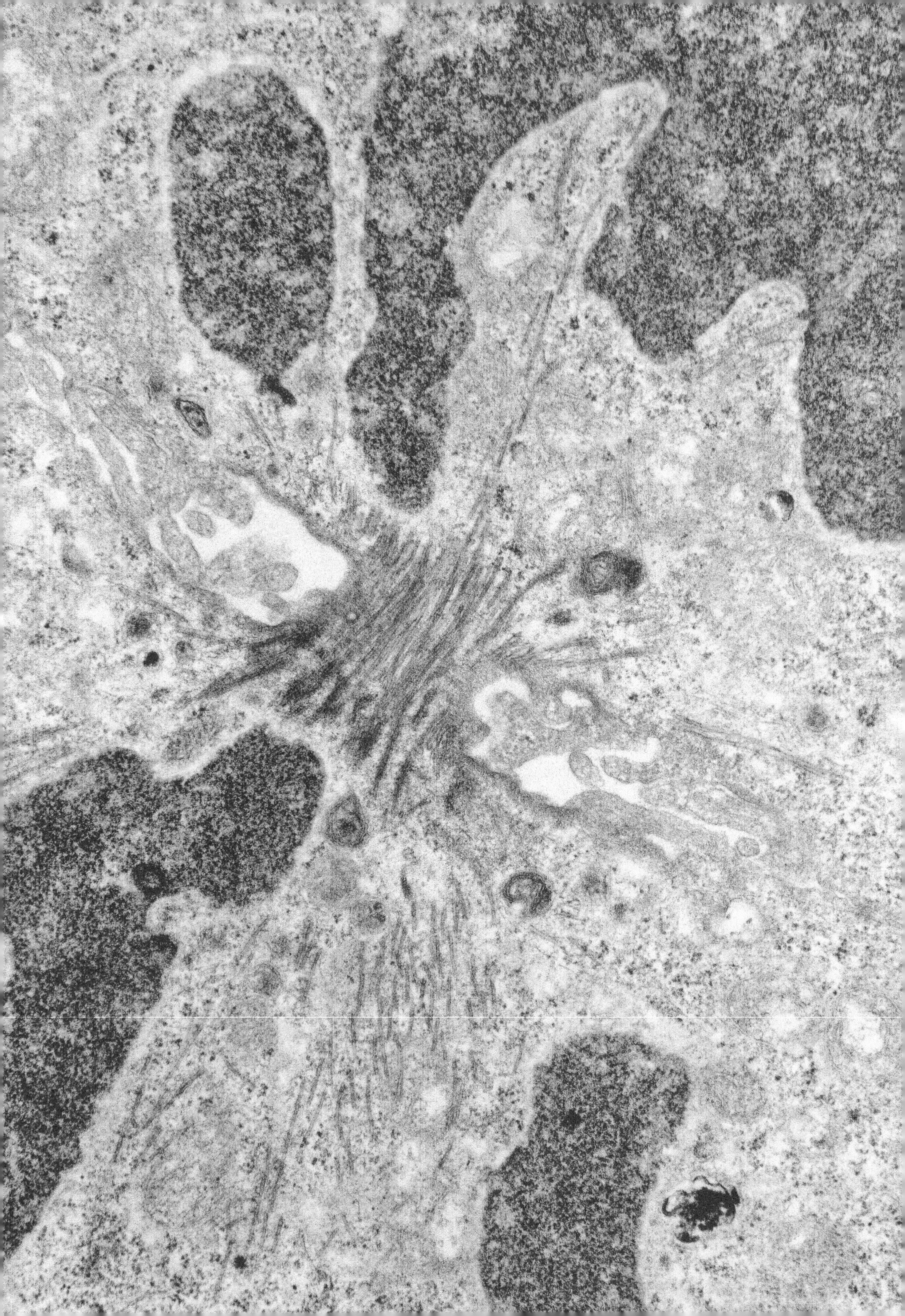

MODULE 1

Cells as the basis of life

Outcomes

By the end of this module you will be able to:

- conduct investigations to collect valid and reliable primary and secondary data and information BIO11-3
- select and process appropriate qualitative and quantitative data and information using a range of appropriate media BIO11-4
- describe single cells as the basis for all life by analysing and explaining cells' ultrastructure and biochemical processes BIO11-8

Content

CELL STRUCTURE

INQUIRY QUESTION **What distinguishes one cell from another?**

By the end of this module you will be able to:

- investigate different cellular structures, including but not limited to:
  - examining a variety of prokaryotic and eukaryotic cells (ACSBL032, ACSBL048) ICT
  - describing a range of technologies that are used to determine a cell's structure and function ICT
- investigate a variety of prokaryotic and eukaryotic cell structures, including but not limited to:
  - drawing scaled diagrams of a variety of cells (ACSBL035) ICT N
  - comparing and contrasting different cell organelles and arrangements CCT
  - modelling the structure and function of the fluid mosaic model of the cell membrane (ACSBL045) CCT ICT

CELL FUNCTION

INQUIRY QUESTION **How do cells coordinate activities within their internal environment and the external environment?**

By the end of this module you will be able to:

- investigate the way in which materials can move into and out of cells, including but not limited to:
  - conducting a practical investigation modelling diffusion and osmosis (ACSBL046) ICT
  - examining the roles of active transport, endocytosis and exocytosis (ACSBL046)
  - relating the exchange of materials across membranes to the surface-area-to-volume ratio, concentration gradients and characteristics of the materials being exchanged (ACSBL047) ICT N

Module 1 • Cells as the basis of life

- investigate cell requirements, including but not limited to:
  - suitable forms of energy, including light energy and chemical energy in complex molecules (ACSBL044)
  - matter, including gases, simple nutrients and ions
  - removal of wastes (ACSBL044)
- investigate the biochemical processes of photosynthesis, cell respiration and the removal of cellular products and wastes in eukaryotic cells (ACSBL049, ACSBL050, ACSBL052, ACSBL053) ICT
- conduct a practical investigation to model the action of enzymes in cells (ACSBL050)
- investigate the effects of the environment on enzyme activity through the collection of primary or secondary data (ACSBL050, ACSBL051) ICT N

Key knowledge

Foundations in biology

Biology is the study of living organisms. **Biodiversity** is the range of living things on Earth. There are close to two million species already identified and many more that remain as yet unidentified.

SCIENTIFIC METHOD

Biologists make observations and construct hypotheses to account for their observations. A **hypothesis** is a possible explanation, an educated guess, made to explain observations.

Hypotheses are tested following the principles of the **scientific method** (Figure 1.1). These include:

- asking relevant questions; that is, questions that can be tested
- making careful observations
- designing and conducting controlled experiments (Figure 1.2). In controlled experiments all **variables** are kept constant, except the one under investigation
- accurate record-keeping of experimental results
- logical interpretation of experimental data and observations
- drawing logical conclusions from the experimental results.

The results of a scientific investigation may negate or reject the hypothesis being tested. In this case the hypothesis must be re-evaluated and modified. Such results are useful in redirecting scientific investigation. When experimental results repetitively support a hypothesis, it may become a **theory** or **principle**. That is, the hypothesis is accepted as a scientific truth.

The scientific method recognises that there are **limiting factors** in investigations. For example, some factors cannot be measured, a sample size may be too small to be representative, or unknown factors may influence investigations.

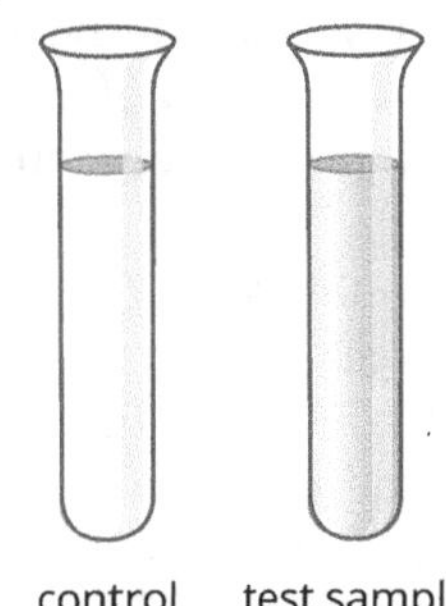

FIGURE 1.2 An experimental set-up

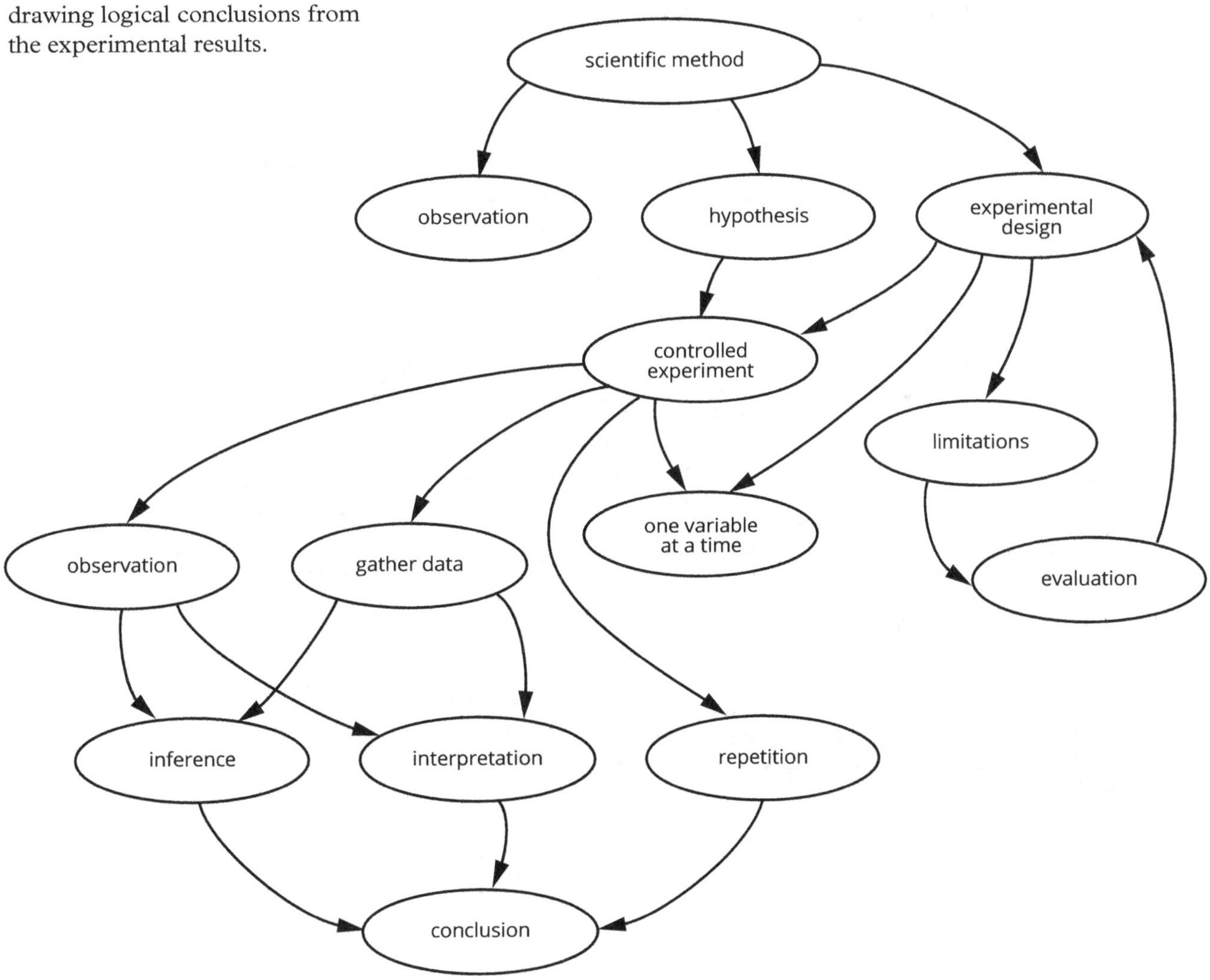

FIGURE 1.1 The scientific method is the process scientists use to conduct experiments and test hypotheses.

Cell structure

CELL THEORY

The cell theory is an important theory in the field of biology. The cell theory states that:

- all organisms are made up of cells
- new cells are produced from existing cells
- the cell is the smallest organisational unit of a living thing.

CELL TYPES

There are two different kinds of cells—**prokaryotic** and **eukaryotic**.

Prokaryotic cells are relatively small and primitive. They do not possess membrane-bound structures. This means they lack sophisticated internal detail. Bacterial cell walls are typically composed of a carbohydrate/protein material called **peptidoglycan** (also known as murein).

While prokaryotic cells do not feature a membrane-bound nucleus, their cells do contain a single, coiled chromosome that contains all of the **deoxyribonucleic acid (DNA)** (genes) necessary to control and direct all the activities of the cell. There are also specialised regions within prokaryotic cells where **cellular respiration** can occur.

Prokaryotes are represented by two domains: **Bacteria** (bacteria and blue-green algae) and **Archaea** (which includes extremophiles). Blue-green algae are photosynthetic bacteria.

Eukaryotic cells (Figure 1.3) are relatively larger and more complex than prokaryotic cells. They possess membrane-bound **organelles** such as a nucleus, mitochondria and lysosomes.

Eukaryotic organisms (Domain Eukarya) include the kingdoms:

- Protista—unicellular organisms
- Fungi
- Plantae
- Animalia.

Note: viruses are non-cellular parasitic agents of disease. They are composed of a core of **ribonucleic acid (RNA)** or DNA surrounded by a protein coat. Prions are also non-cellular agents of disease, but they are composed only of protein.

CELL ORGANELLES

Organelles are distinct structures within cells that perform specific functions. Some organelles are visible using the light microscope, while others are not. The details of many cell organelles are only visible when using the electron microscope.

Some key organelles are summarised in Figure 1.3 and Table 1.1.

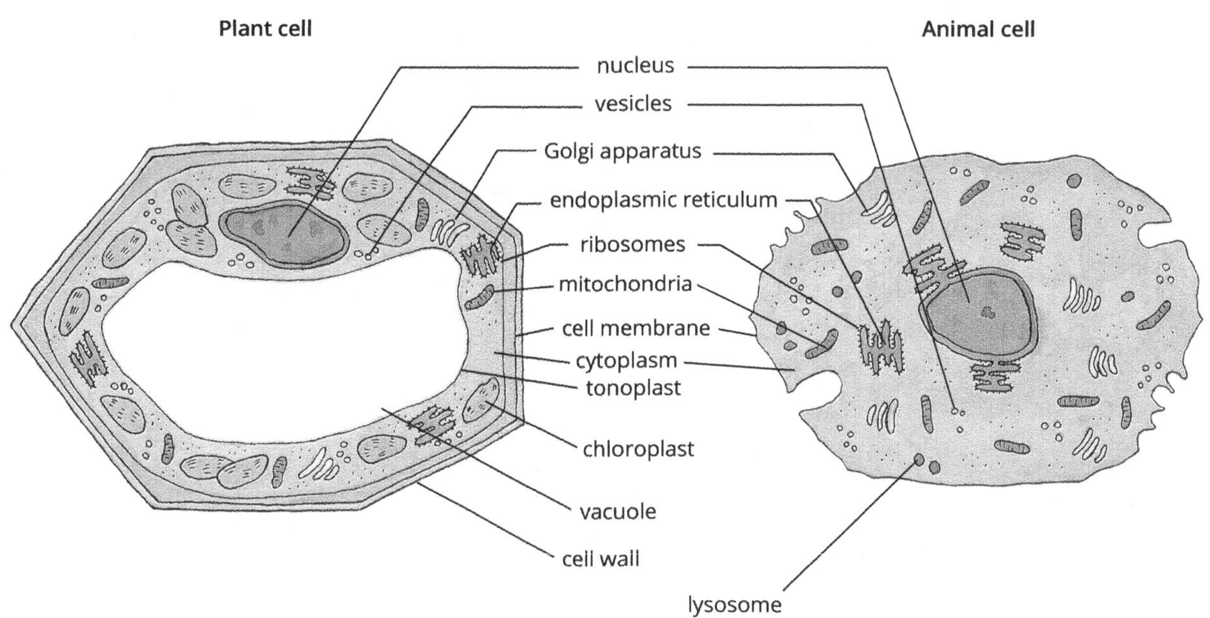

FIGURE 1.3 In unicellular and multicellular organisms, many of the functions that are essential to life occur within specialised structures and organelles of individual cells.

 ISBN 978 1 4886 1931 1

TABLE 1.1 Cellular organelles and their functions

Organelle	Description and function	Found in both plants and animals
Nucleus	large spherical organelle; controls cell activities (contains DNA)	yes
Mitochondrion	features folded inner membrane; site of aerobic stages of cellular respiration (contains some DNA)	yes
Ribosomes	tiny spherical organelles; site of protein synthesis; not membrane-bound	yes
Endoplasmic reticulum (ER)	network of membranes involved in protein transport within cells; ER encrusted with ribosomes is called 'rough' ER	yes
Golgi apparatus	stacks of flattened membranous sacs; modifies and packages substances in preparation for secretion from cell	yes
Chloroplast	site of photosynthesis (contains chlorophyll)	plant cells only
Lysosomes	membrane-bound organelles that produce digestive enzymes; break down complex compounds into simpler molecules	animals plants (some evidence)
Vacuoles	membrane-bound compartments that keep a variety of substances separate from cell contents (large in plant cells, small in animal cells)	yes
Cilia*	short and hair-like; generally present in large numbers; rhythmic waves create movement of substances over cell surface, or movement of the cell	yes
Flagellum*	long and hair-like; generally singular or present in small numbers; rhythmic contractions enable movement of cell	yes
Cell wall	rigid structure surrounding cell; composed of cellulose in plants limits cell expansion when fully turgid; contributes to structural support of plant	plant cells only
Cell membrane	semipermeable, flexible barrier; controls cell inputs and outputs	yes

* Present in some cells only.

Endosymbiotic theory

Mitochondria and chloroplasts are organelles that display some unusual features, including their own ribosomes and DNA in the form of a single circular chromosome. They also feature a double membrane and the ability to replicate independently of the cells that contain them. Scientists believe these prokaryotic-like features suggest they originated around 1.5 billion years ago as free-living bacteria that were engulfed by other free-living bacteria in a relationship that offered mutual advantage. This is called the **endosymbiotic theory**.

INVESTIGATING CELLS

Cytology is the study of cells. Cytologists use a variety of tools and techniques to study cells. The main tool used by cytologists is the microscope. There are many different types of microscopes but the two main ones are the light microscope and the electron microscope.

TABLE 1.2 Comparison of microscopes

Feature	Light microscope	Electron microscope
Magnifying power	low	high
Cost	low	high
Level of expertise needed	low	high
Can living specimens be observed?	yes	no

Cell size

Cells vary greatly in size. Most cells are only visible under a light microscope, and their size is usually measured in micrometres (μm). There are 1000 micrometres in 1 millimetre (mm). Although most cells are microscopic, there are some exceptions—the egg cell of some bird species can be many centimetres (cm) in diameter.

Some typical cell sizes are:

- bacterium: 0.1–1.5 μm long
- human: 8–60 μm long
- plant: 10–100 μm long
- paramecium (a single-celled eukaryote): about 150 μm long.

The thickness of cell membranes also differs between cells, and can be between 0.004 and 0.1 μm thick.

The light microscope

Most cells are so small that they can only be seen with a microscope. The light microscope (also called the compound microscope, Figure 1.4) uses light and a system of lenses to magnify the image. One lens is called the objective lens and the other is the eyepiece or ocular lens. The total magnification of a microscope is calculated by multiplying the magnifying power of the ocular lens (eyepiece) by the magnifying power of the objective lens. For example, an ocular lens with a magnifying power of 4 times (×4) used with an objective lens with a magnifying power of 10 times (×10) gives a total magnification of 40 times (×40).

Beneath the microscope stage is a mirror or built-in light source, a condenser lens system to concentrate the light, and an iris diaphragm mechanism to regulate the amount of light passing through the object.

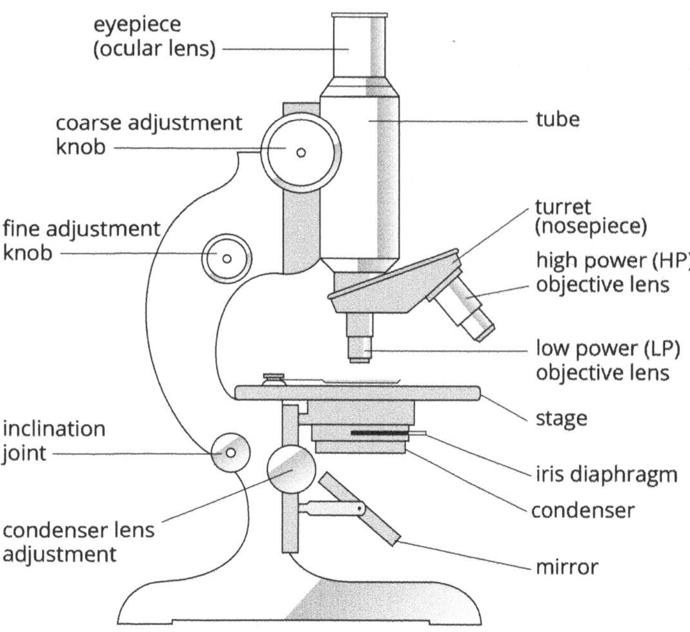

FIGURE 1.4 A light microscope

User-friendly microscope hints

- Look down the eyepiece (ocular lens) and adjust the light source (mirror, condenser lens and iris diaphragm) so the field of view is uniformly illuminated.
- Place your prepared slide on the microscope stage and centre the object to be viewed. Use the clips to secure the slide in position.
- When setting up the microscope always view the object under low power (LP) first (usually ×10).
- Checking from the side, wind the coarse adjustment knob until the LP objective lens is as close as it can go towards the slide (it should be no closer than 2 mm).
- Looking down through the eyepiece, use the coarse adjustment knob to slowly move the LP objective lens away from the slide. When the object is in focus, use the fine adjustment knob to bring the image into even sharper focus.
- Rotate the turret to set a high power (HP) objective lens (×40) in place. Only use the fine adjustment knob when using HP.
- When using the ×100 objective lens, the clarity of the image can be improved by using a technique called oil immersion. Oil immersion involves adding a drop of oil to the coverslip of the slide and carefully rotating the turret so that the ×100 objective lens comes into contact with the drop of oil. Only the fine adjustment knob should be used when using oil immersion.

Preparing a wet mount slide

Figure 1.5 illustrates the best technique for making a wet mount slide.

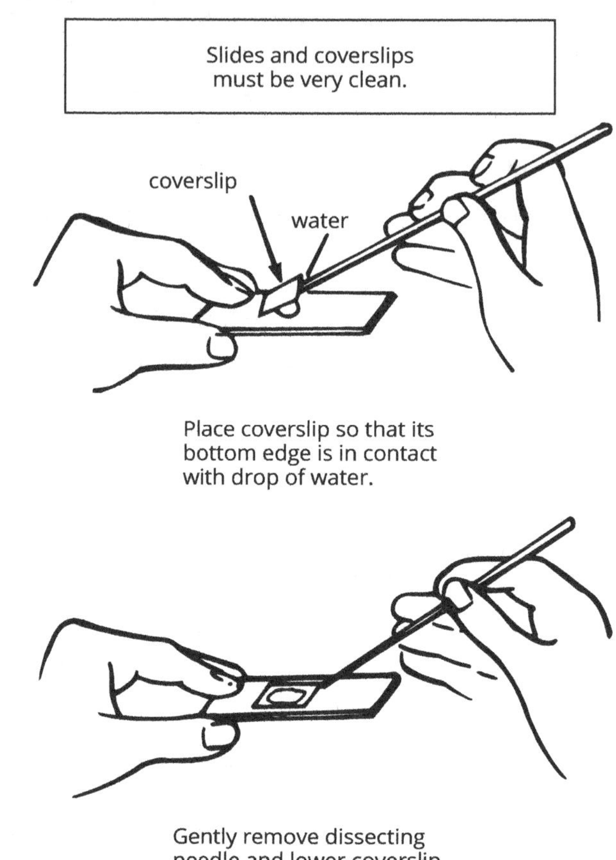

FIGURE 1.5 Making a wet mount slide

Biological drawings

The following guidelines should help you to make simple and effective scaled diagrams of biological structures.

- Drawings should be:
  - made in grey lead pencil
  - large
  - fully labelled with the name of the specimen, the type of preparation and the magnification
  - given a size perspective so that comparisons can be made between specimen sizes—draw each specimen in relation to the size of the field of view observed.
- Lines to labels should be ruled—they should not have 'arrowheads' and should not cross over.
- Drawings of low-power images should not show the detail of cells, just the 'area of cell types' (Figure 1.6).
- Drawings of images made under high power should show the detail of a few cells only of each type (Figure 1.7).

ISBN 978 1 4886 1931 1

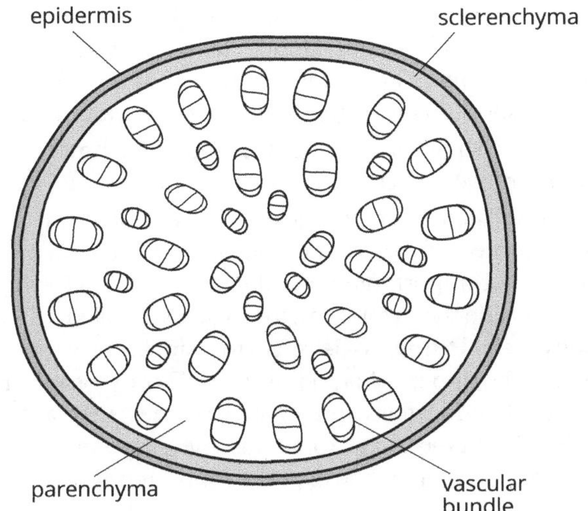

FIGURE 1.6 A low-power view of a zea maize stem (×30)

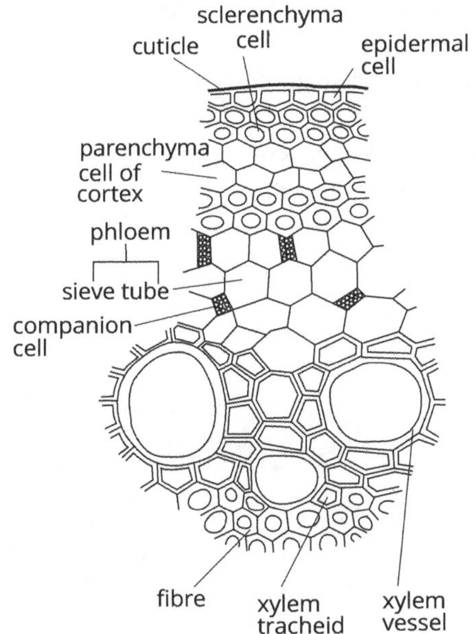

FIGURE 1.7 A high-power view of zea maize cells (×600)

Cell function

MOVEMENT OF MATERIALS IN AND OUT OF CELLS

The **internal environment** of cells is the **intracellular fluid**—the medium inside cells. The **external environment** of cells is the **extracellular fluid**—the watery medium surrounding cells.

Cell membranes

The **cell membrane** (also known as the plasma membrane) controls entry and exit of substances into and out of cells (Figure 1.8). It controls which substances leave and enter, when and how much (Table 1.3). It responds to instructions from the nucleus. It can detect and respond to external stimuli.

The cell membrane is described as being **semipermeable** (also called partially permeable) because it is permeable to some substances but not others.

The composition of the cell membrane is basically the same as that of all membranes within cells (including the membranes of the nuclear envelope, mitochondria, Golgi apparatus, endoplasmic reticulum, vacuoles, lysosomes and chloroplasts).

The cell membrane consists of a double layer of special lipid molecules called **phospholipids**. This is called the phospholipid bilayer. The bilayer has protein molecules scattered through it in a random arrangement. The total structure is fluid. This means that the molecules can move around relative to each other. The structure of the cell membrane is commonly described using the **fluid mosaic model**.

In summary, the cell membrane is a flexible, semipermeable barrier between the intracellular and extracellular environments.

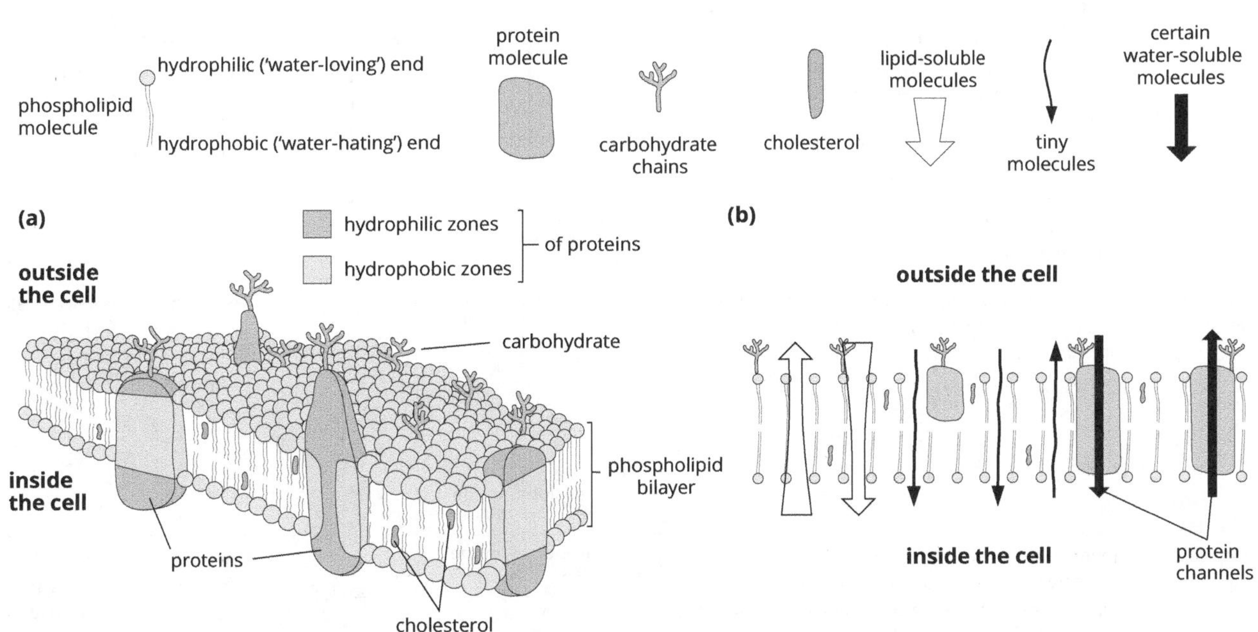

FIGURE 1.8 (a) Biological membranes are composed of a phospholipid bilayer with large protein molecules embedded in the bilayer. These proteins provide channels for the passive and active movement of certain molecules across the cell membrane. (b) Short carbohydrate molecules attached to the outside of the membrane are involved in cell adhesion and cell recognition.

Movement across the cell membrane

The exchange of materials between cells and their external environment occurs through the processes of:

- **diffusion**
- **osmosis**
- **facilitated diffusion,** or
- **active transport**.

The process by which substances move across the cell membrane depends on several factors—these include the lipid nature of the cell membrane and the size and polarity of molecules. **Hydrophilic** substances dissolve in water and do not readily pass across phospholipid membranes. **Hydrophobic** substances do not readily dissolve in water—such molecules can dissolve in the phospholipid membrane.

- Lipid-soluble molecules (non-polar) such as chloroform and alcohol dissolve in the lipid bilayer and pass through.
- Water-soluble molecules tend to be repelled by the phospholipid bilayer. However, very small molecules such as water and urea are small enough to pass directly between the phospholipid molecules.
- Other small uncharged molecules such as oxygen and carbon dioxide also pass directly across the cell membrane between the phospholipid molecules.
- Large water-soluble molecules (polar) such as simple sugars and amino acids cannot pass directly across the cell membrane. The passage of these molecules depends on transport channels that span the cell membrane.

Table 1.3 shows the factors that influence how materials move across the cell membrane.

Some substances enter and leave cells by other means. When a section of the cell membrane wraps around a substance for import into the cell, pinching off to form a vesicle inside the cytoplasm, the process is called **endocytosis**. **Pinocytosis** refers to a similar process related to the import of liquid droplets. **Exocytosis** is the opposite of endocytosis and involves vesicles, such as those associated with the Golgi apparatus, merging with the cell membrane to facilitate the export of substances.

Surface-area-to-volume ratio

When substances enter or leave cells, the rate at which they move is determined by a number of factors. These include:

- concentration (a steep concentration gradient causes faster diffusion)
- temperature (higher temperatures increase the rate of movement of molecules)
- surface-area-to-volume ratio (SA : V).

TABLE 1.3 Movement across the cell membrane

Process	Description	Active or passive	Diagram	Example in organism
Diffusion	movement of particles from an area of high concentration to an area of low concentration, along a concentration gradient	passive (does not require energy)		oxygen enters body cells (low in O_2 since continually using O_2 in cellular respiration) from the capillaries where it is in high concentration (O_2 replenished at lungs)
Osmosis	special type of diffusion that involves movement of water molecules across a semipermeable membrane; water moves from an area of high concentration of free water molecules to an area of low concentration of free water molecules, i.e. low solute concentration to high solute concentration	passive	starch molecule; water molecule	cells in kidney medulla absorb water by osmosis due to osmotic gradient between ion concentration in tissue fluid and the kidney tubules
Facilitated diffusion	movement of particles from high to low concentration through a protein channel in cell membrane	passive	protein channel; outside cell; inside cell; amino acid	small molecules such as amino acids and glucose enter cell via protein channel
Active transport	movement of particles from an area of relatively low concentration to an area of high concentration, against a concentration gradient	active (requires input of energy)	outside cell; inside cell; ion	uptake of ions by root hair cells of plants and uptake of nutrients by gut epithelium cells of animals, so that concentration within cells exceeds concentration in external medium

ISBN 978 1 4886 1931 1

Consider the two cells in Figure 1.9. Which cell has the larger SA : V? The answer is Cell B. Although Cell A has a larger volume, Cell B has a larger surface area compared to its volume (a larger surface-area-to-volume ratio). This means it will be more efficient at taking in and exporting substances through its cell membrane per unit time.

In general, the surface-area-to-volume ratio of an organism decreases as size increases. Cells and organisms have structural adaptations to overcome this. Such adaptations include microvilli on absorptive cells, and the ribbon-like body shape of tapeworms.

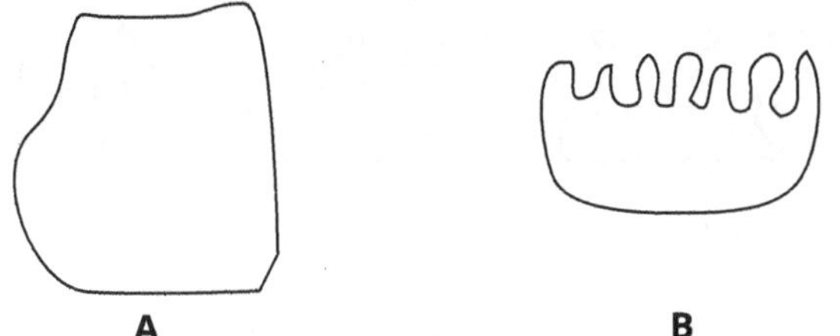

FIGURE 1.9 Two cells (A and B) with a different volume but similar surface area

CELL REQUIREMENTS

The cells of living organisms share some basic requirements to ensure their survival. These include:

- water—the cytoplasm of cells is largely composed of water; cells also need a watery external environment
- oxygen—for cellular respiration (with the exception of some organisms, such as anaerobic bacteria)
- inorganic nutrients (for example, minerals).

The photosynthetic cells of plants also require an input of carbon dioxide.

Cells also need optimal conditions to maximise their efficiency. This can be different for the cells of different kinds of organisms; for example, the temperature requirements of cells are different for animals such as mammals and fish.

All cells have in common a need to access nutrients and dispose of wastes. The semipermeable membranes of cells allow for this. In general, the features of cells, including their structure, are related to their function. For example, photosynthetic plant cells contain **chloroplasts**, the site of **photosynthesis**. This specialisation of cells is considered further in Module 2.

Inorganic and organic materials

Cells are composed of and require chemicals to function. The main molecule found in cells is water. Some plant cells are more than 90 per cent water. In addition to water, cells consist of both inorganic and organic substances.

Inorganic compounds (including water) are relatively simple and do not contain hydrocarbon groups. **Organic compounds** are relatively complex and contain hydrocarbon groups.

Table 1.4 provides a summary of the inorganic and organic compounds that are required by cells.

TABLE 1.4 Inorganic and organic compounds required by cells

Substance	Examples	Function(s) in cells
Inorganic compounds		
Water	H_2O	All chemical reactions in organisms take place in solution in water. Water has high heat capacity.
Oxygen	O_2	Oxygen is needed for efficient energy supply, achieved by the process of cellular respiration in almost all organisms. It is taken in as a gas by terrestrial organisms and in solution by aquatic ones.
Carbon dioxide	CO_2	Carbon dioxide is the main source of the carbon atoms for organic molecules, usually starting with carbon fixation by photosynthesis in autotrophs (green plant cells). CO_2 is taken into plant leaves as a gas, converted to sugars and eventually returned to the atmosphere in the carbon cycle.
Minerals	nitrogen (N) phosphorus (P) iron (Fe) magnesium (Mg)	N is used for protein and nucleic acid synthesis. P is used for nucleic acid synthesis and is an important component of cell membranes. Fe is a component of haemoglobin in red blood cells. Mg is a component of chlorophyll.
Organic compounds		
Carbohydrates	basic building blocks; are monosaccharides; contain C, H, O, N	Carbohydrates provide an energy source to cells that can be accessed relatively easily.
Lipids	basic building blocks; are glycerol and fatty acids; contain C, H, O	Lipids are used for long-term energy storage and insulation, and are structural components of membranes.
Proteins	basic building blocks; contain C, H, O, N	All enzymes are proteins. Proteins also play important structural roles.
Nucleic acids	in DNA and RNA; contain C, H, O, N, P	DNA carries the genetic code. RNA is involved in transcription and translation of the genetic code.
Vitamins	vitamin C vitamin D	Vitamin C prevents scurvy. Vitamin D facilitates the uptake of calcium into bones. Bone vitamins have important roles in enzyme function, for example as coenzymes.

BIOCHEMICAL PROCESSES IN CELLS

Some cell functions have already been listed and described above. Cell functioning relies on chemical reactions that occur within and between cells. The chemical reactions that take place in an organism are known as biochemical processes or **metabolism**. Metabolism can be divided into two main types of processes:

- **endergonic** processes that require a net input or use of energy
- **exergonic** processes that result in a net output or release of energy.

Processes in which energy is changed from one form to another are energy transformations. Photosynthesis and cellular respiration are examples of energy transformations that occur in living organisms.

- Photosynthesis is an example of an endergonic process.
- Cellular respiration is an example of an exergonic process.

Photosynthesis

Photosynthesis is performed by plant cells that contain **chlorophyll** (Figure 1.10). It requires light and involves the conversion of light energy into chemical energy.

A word equation for photosynthesis is as follows.

$$\text{carbon dioxide} + \text{water} \xrightarrow[\text{chlorophyll}]{\text{light}} \text{glucose} + \text{oxygen}$$

A simple balanced equation for photosynthesis is:

$$6CO_2 + 6H_2O \xrightarrow[\text{chlorophyll}]{\text{light}} C_6H_{12}O_6 + 6O_2$$

We can also represent photosynthesis in the following way, as water is in fact an input and an output for this process.

$$6CO_2 + 12H_2O \xrightarrow[\text{chlorophyll}]{\text{light}} C_6H_{12}O_6 + 6O_2 + 6H_2O$$

Organisms that produce their own organic compounds are called **autotrophs**. Autotrophs can be photosynthetic (plants) or chemosynthetic (some prokaryotes) (Figure 1.11). Organisms such as fungi and animals that obtain their organic compounds from other organisms are called **heterotrophs**.

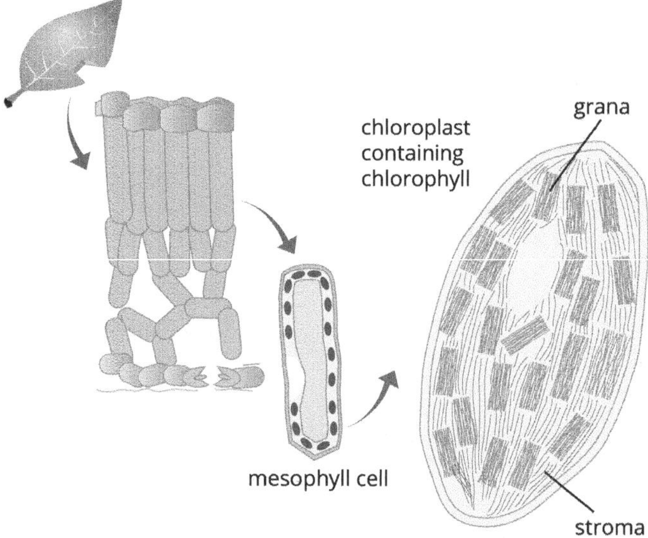

FIGURE 1.10 Leaves, and some stems, are green because the mesophyll cells contain many chloroplasts (the organelles in which photosynthesis takes place).

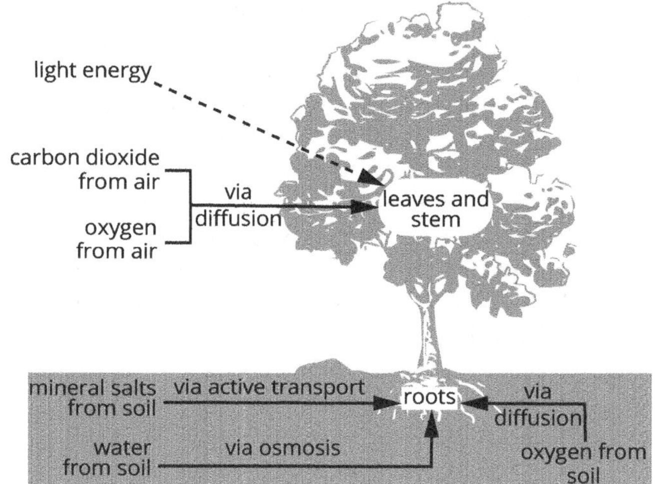

FIGURE 1.11 Overview of how plants obtain nutrients

Photosynthesis occurs in two stages: the **light-dependent reaction** occurs in **grana**; the **light-independent reaction** occurs in the **stroma** (Table 1.5).

Adaptations such as smaller leaves and orientation of leaves can reduce water loss. Some plants are adapted to hot, dry conditions by closing **stomata** during the day and opening them at night to take up the carbon dioxide needed for photosynthesis.

Factors that may limit the rate of photosynthesis include light intensity, carbon dioxide concentration and temperature. Photosynthesis and cellular respiration occur simultaneously in green plants during periods of light exposure. The point at which the products of photosynthesis are consumed in the process of cellular respiration is called the **light compensation point**.

TABLE 1.5 Overview of stages involved in photosynthesis

1st stage	2nd stage
Light-dependent	Light-independent

• occurs in grana
• red and blue light absorbed
• light absorbed by chlorophyll
• energy used to split water molecules
• creates O_2 (by-product) and H^+ ions
• ATP also produced (used in 2nd stage)
	• occurs in stroma
	• carbohydrate produced
	  - in the form of glucose
	  - stored as starch
	• H^+ ions and CO_2 (from air) combined
	• ATP from 1st stage consumed in glucose manufacture

Cellular respiration

Adenosine triphosphate (ATP) is the immediate source of energy for cells. It is produced in a series of chemical reactions that involves the breakdown of organic molecules. The useable energy of ATP is contained in the phosphate bonds of the molecule. The cycling of ATP and **adenosine diphosphate (ADP)** means that energy continues to be available for use in the cell (Figure 1.12).

Cells access the energy available in organic molecules through **glycolysis** (anaerobic), and either cellular respiration (aerobic) or **fermentation** (anaerobic).

ISBN 978 1 4886 1931 1

Cellular respiration is the process in which complex organic compounds are broken down to release energy (ATP). Water is a by-product.

A word equation for aerobic cellular respiration is as follows:

glucose + oxygen → carbon dioxide + water + energy

A balanced chemical equation for cellular respiration is:

$$C_6H_{12}O_6 + 6O_2 \longrightarrow 6CO_2 + 6H_2O + \text{energy}$$

36 (ADP + Pi) → 36 ATP

glucose
ATP
energy for work
energy
carbon dioxide and water
ADP
adenosine
phosphate
energy-containing bond

FIGURE 1.12 ATP/ADP cycle

Anaerobic respiration

If oxygen is not available to meet the cell's energy requirements, **anaerobic respiration** will occur (fermentation). Lactic acid is produced in animal cells. Ethanol and carbon dioxide are produced in plant and yeast cells. Less ATP is produced during anaerobic respiration. Anaerobic respiration is less efficient than aerobic respiration.

ENZYME ACTIVITY IN CELLS

The metabolism of an organism is the sum of all the chemical reactions that occur within its cells. This includes the energy-transforming reactions of cells such as production of organic molecules, and the breakdown, recycling and excretory processes. These biochemical processes are universal, they occur in the cells of all living organisms to ensure the survival of the individual.

Enzymes are biological **catalysts**—they increase the rate of biochemical reactions in cells, for example the chemical reactions involved in cellular respiration and photosynthesis.

Properties of enzymes

Enzymes are usually denoted by the suffix 'ase', for example maltase, lactase, protease, amylase and lipase.

Enzymes:

- are composed of protein
- are substrate specific—they catalyse a chemical reaction involving a particular substrate molecule, and not any other. There are two theories used to explain how enzymes interact with their substrates: **lock-and-key**, and **induced fit** (Figures 1.13a and b).

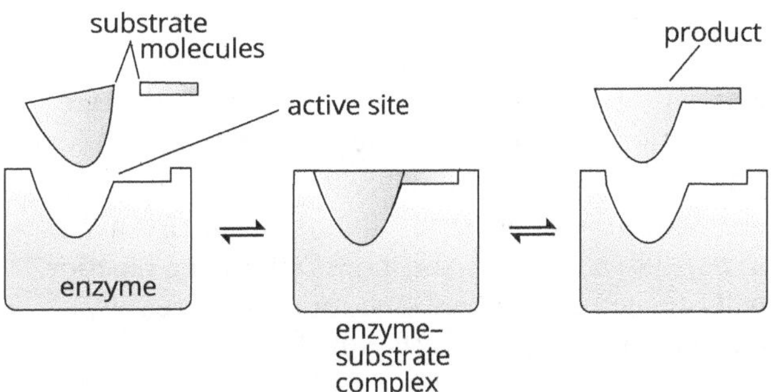

FIGURE 1.13a Model of enzyme 'lock-and-key' operation

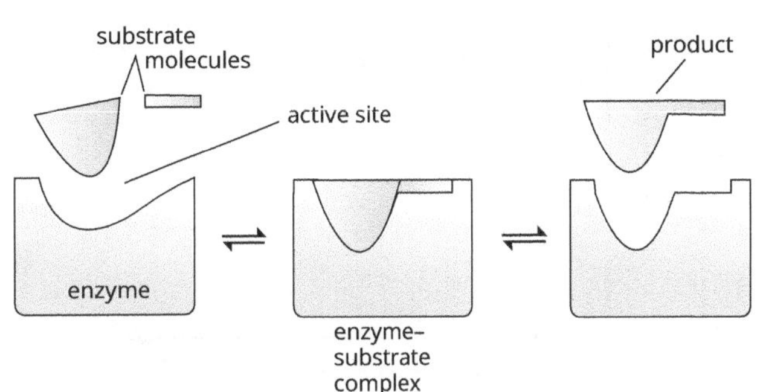

FIGURE 1.13b Model of enzyme 'induced fit' operation

- take part in chemical reactions but are not used up or changed by the process. They are released at the end of a reaction and so are available to be used over and over again
- have optimal conditions under which they work most effectively; they will catalyse a reaction so that maximum product is produced per unit of time. For example, the optimal conditions for digestive enzymes in the duodenum of humans is a temperature of 37°C and pH 8
- are sensitive to factors such as temperature and pH. When these conditions are not optimal, the activity of enzymes is reduced. Extremes of such factors may lead to enzymes becoming **denatured**. When this happens the enzyme cannot recover its function because the shape of its **active site** has been permanently altered (Figure 1.14).

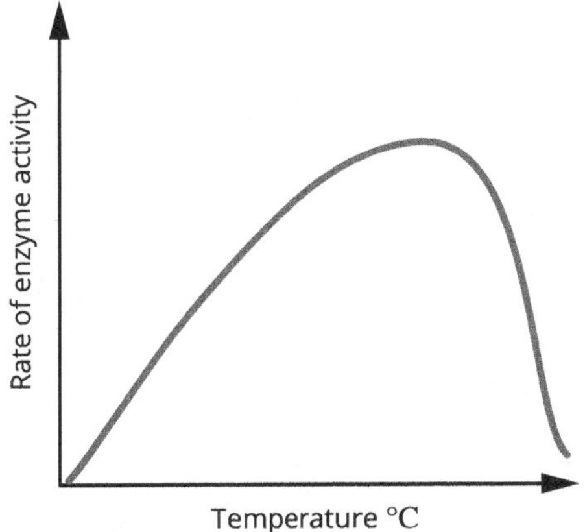

FIGURE 1.14 Rate of enzyme activity continues to increase with increasing temperature. Excessive temperature denatures enzymes and activity ceases.

The rate of enzyme activity is dependent upon the:

- concentration of substrate—the higher the concentration of the substrate, the greater the rate of interaction between substrate molecules and enzymes, leading to an increased rate of reaction
- concentration of enzyme—the more enzyme available to catalyse a reaction, the more rapidly the reaction will proceed until all enzyme molecules are fully engaged in the reaction (Figure 1.15).

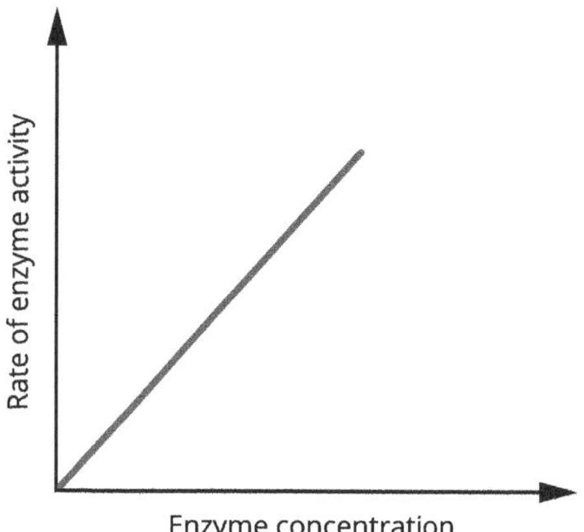

FIGURE 1.15 Rate of enzyme activity increases with increasing enzyme concentration

Reversible and irreversible enzyme inhibition

The action of enzymes is also influenced by the presence of chemical competitors at the active site. Competitors inhibit enzyme action by binding to the active site of enzymes. This may be reversible or irreversible.

Penicillin is an example of an irreversible enzyme inhibitor. Strong covalent bonds means the competitor is permanently attached to the active site of an enzyme involved in construction of the bacterial wall. This outcome renders pathogenic bacteria unable to survive, hence the use of penicillin in medicine.

Reversible inhibition occurs when molecules form weak, non-permanent bonds with the active site of enzymes. Reversible enzyme inhibition has an important regulatory role in cells. Inhibitors are in place when enzymes are not required and dissociate from the enzyme when the enzyme product is needed.

- **Activation energy**: the energy expended to initiate a reaction; even catabolic reactions require an initial input of energy to start the reaction. Enzymes lower the activation energy, making it easier for the reaction to proceed.
- **Coenzymes**: small organic molecules important to the normal functioning of enzymes.
- **Cofactors**: inorganic ions important to the normal functioning of enzymes, for example Mg++.

 ISBN 978 1 4886 1931 1

WORKSHEET 1.1

Knowledge review—basic biology

This activity addresses foundation ideas in biology that you have studied before and on which the key ideas in this first module are built. Follow the instructions to complete these introductory summaries. Remember to be resourceful, referring to text material or class notes as needed.

	Instruction	Response	
1	Draw and label a typical plant cell to show the basic similarities and differences as compared to the typical animal cell.	**Animal cell** cytoplasm cell membrane nucleus	**Plant cell**
2	The prefix 'pro' comes from the Greek word for 'before' and also translates as 'early' or 'primitive'. 'Eu' stems from the Greek word for 'true'. 'Karyon' is Greek for 'nut' and is used to describe the shape of the nucleus in certain cells. Use this information to complete the description of eukaryotic organisms and how they compare to prokaryotic organisms already described. Include a labelled diagram.	**Prokaryotic organisms** Prokaryotic organisms have cells that are characterised by the lack of a distinct, membrane-bound nucleus. Bacteria are prokaryotic organisms. bacterial chromosome cell membrane cell wall cytoplasm	**Eukaryotic organisms**
3	All organisms have specific nutritional requirements that must be met so they can survive. Consider the list of requirements shown for animals and complete the list for typical plants.	**Animal requirements:** • oxygen (for cellular respiration) • water • carbohydrates • proteins • lipids • vitamins • minerals	**Plant requirements:**
4	In processing nutrients in cell metabolism, organisms produce waste materials that are toxic and must be removed before they can cause harm. Some waste materials for mammals are listed. For each waste material, identify the organ that is involved in its removal.	**Waste material:** • carbon dioxide • urea (nitrogenous waste produced by cells) • excess water • excess salt	**Organ:**
5	Cellular respiration and photosynthesis are biochemical processes that result in specific energy transformations in different kinds of organisms. These processes are critical to the survival of the organisms in which they occur. Define each process. Include the raw materials for each and the substances produced. Look carefully at the information for each process. Summarise the pattern or relationship you notice between these two processes.	**Cellular respiration**	**Photosynthesis**

WORKSHEET 1.2

Poor pot plant—controlled experiments

A biology student designed and conducted the following experiment to test a hypothesis. Read the student's experimental procedure and cast a critical eye over the results obtained.

Purpose

To investigate the effect of sunlight on green plants.

Hypothesis

If green plants need sunlight to survive, then plants that do not receive sunlight will die.

Procedure

The student:

1 obtained two seedlings of the same species that were the same in other respects, including size, height and weight

2 labelled two same-size pots 'Pot plant 1' and 'Pot plant 2'

3 potted the seedlings using a commercial potting mix, but ran out of potting mix for Pot plant 2 and topped it up with some soil from the school garden

4 gave both plants 100 mL of water

5 placed Pot plant 1 on the window sill with plenty of exposure to sunlight and Pot plant 2 in a dark cupboard below the sill to ensure it had no access to sunlight.

Pot plant 1

Pot plant 2

The student watered Pot plant 1 on the window sill every two days for a period of two weeks but forgot about Pot plant 2, which was out of sight in the cupboard. At the end of the two-week period, Pot plant 1 was thriving but Pot plant 2 was dead.

Conclusion

The student wrote the following conclusion:

Pot plant 2 was not exposed to sunlight and it died. The plant must have died due to lack of sunlight. The hypothesis that 'green plants need sunlight to survive' is supported by the results of the experiment.

1 Which pot plant represented the control in this experiment? Explain your choice.

2 a How many variables did the student include in the experiment? What were they?

b How many factors should be varied in a properly controlled experiment? What should the variable be in this instance?

ISBN 978 1 4886 1931 1

3 Are the student's conclusions accurate? Explain.

4 Outline the conclusions that could be drawn from the student's experiment.

5 **a** Describe the changes you would make to ensure this is a properly controlled experiment.

b In light of the changes you would make, outline the results you would expect and the conclusions you could draw.

RATING MY LEARNING	My understanding improved	Not confident ◄──► Very confident ○ ○ ○ ○ ○	I answered questions without help	Not confident ◄──► Very confident ○ ○ ○ ○ ○	I corrected my errors without help	Not confident ◄──► Very confident ○ ○ ○ ○ ○

WORKSHEET 1.3

The inside story on cell structure

1 Study the following diagrams of plant and animal cells. Identify the organelles indicated in the spaces provided.

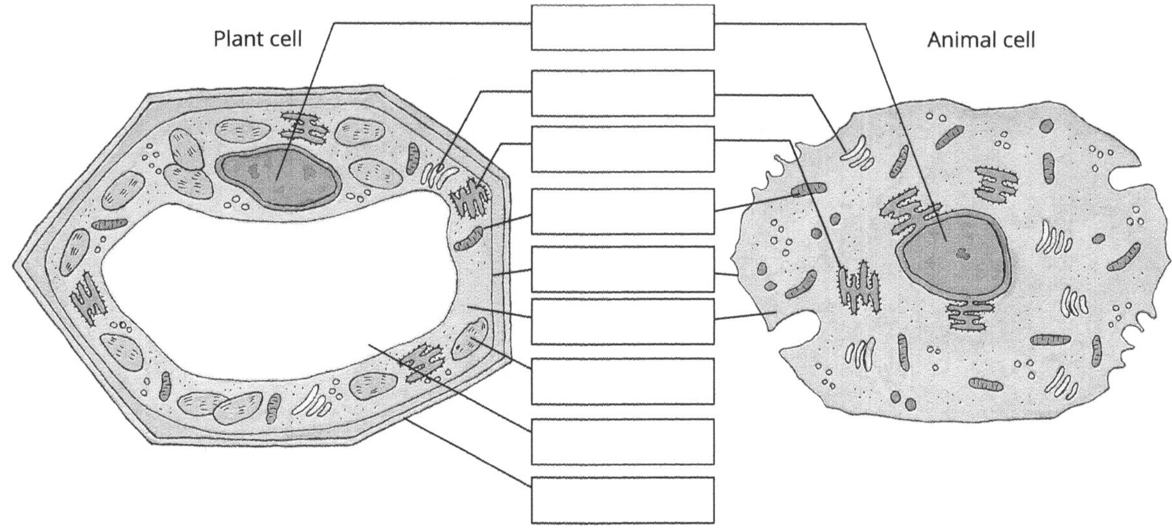

2 Examine the electron micrographs of each of the different cell organelles shown in the following table. Use the diagrams above to help you identify each. Complete the details in the table to provide a summary of the structure and function of each organelle.

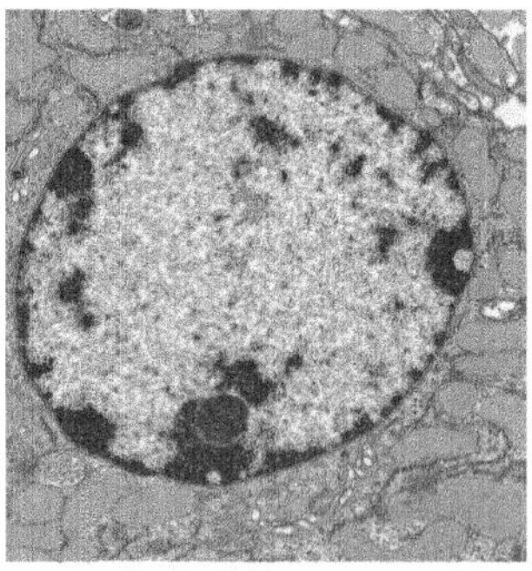

	Name of organelle: nucleus **Description:** **Function:**	A B	**Name of organelle A:** rough endoplasmic reticulum **Description:** **Function:** **Name of organelle B:** ribosome **Description:** **Function:**
	Name of organelle: **Description:** **Function:**		
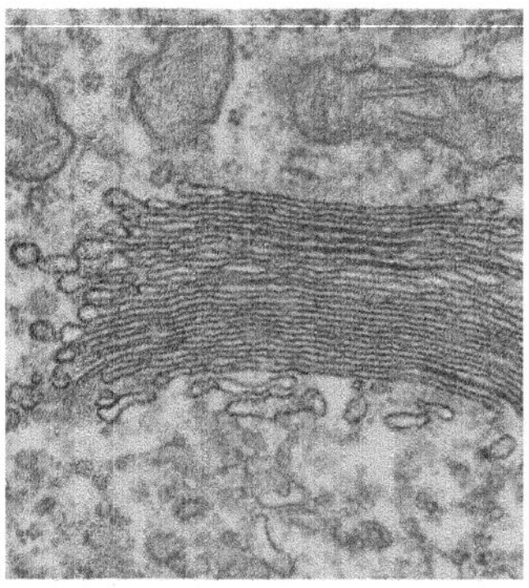	**Name of organelle:** Golgi apparatus **Description:** **Function:**	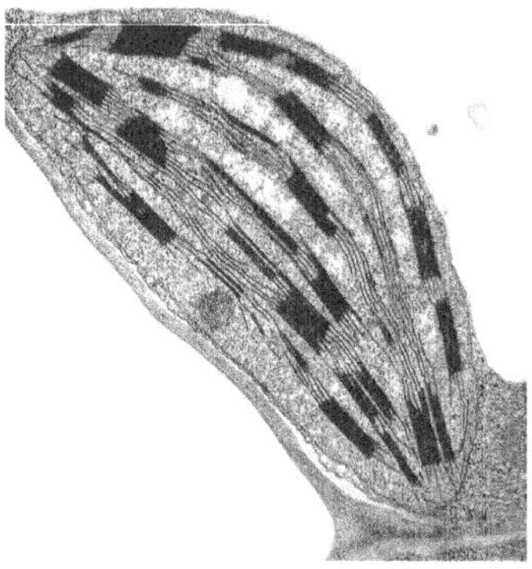	**Name of organelle:** chloroplast **Description:** **Function:**

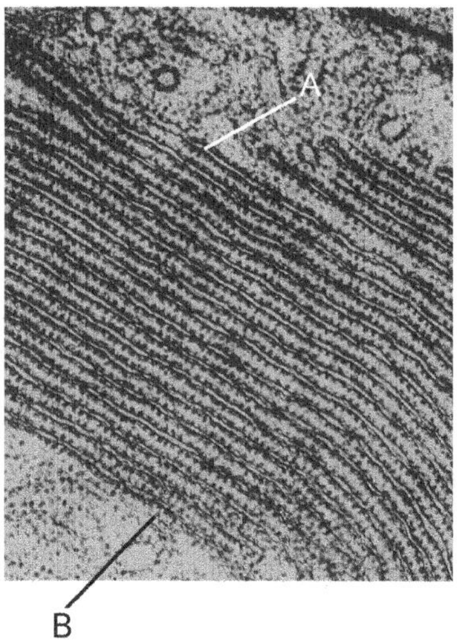

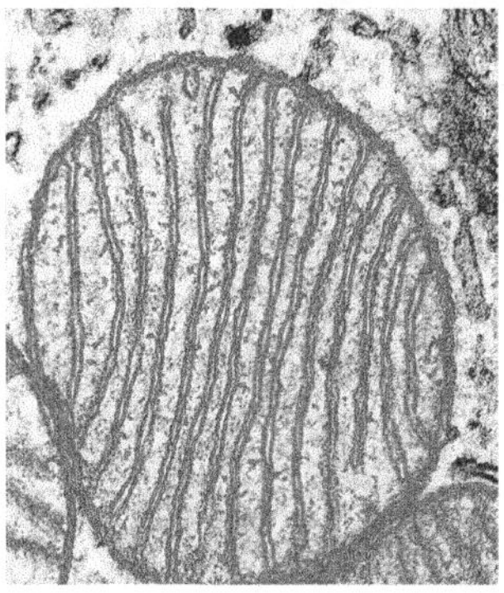

 ISBN 978 1 4886 1931 1

3 a The following organelles are shown in question 1 above. Describe the role of each in the table.

Organelle	Role
cell membrane	
cytoplasm	
cell wall	
vacuole	

b The following organelles are not shown in question 1 above. Use your textbook or other resources to describe each organelle and outline their roles.

Organelle	Description	Role
cytoskeleton		
centriole		

4 a Describe two similarities between plant and animal cells.

b Describe two differences between plant and animal cells.

RATING MY LEARNING	My understanding improved	Not confident ◄——► Very confident ○ ○ ○ ○ ○	I answered questions without help	Not confident ◄——► Very confident ○ ○ ○ ○ ○	I corrected my errors without help	Not confident ◄——► Very confident ○ ○ ○ ○ ○

WORKSHEET 1.4

Model membranes—structure and function

1 The composition of the cell membrane is commonly described using the fluid mosaic model. Complete the diagram by adding labels and functions where indicated.

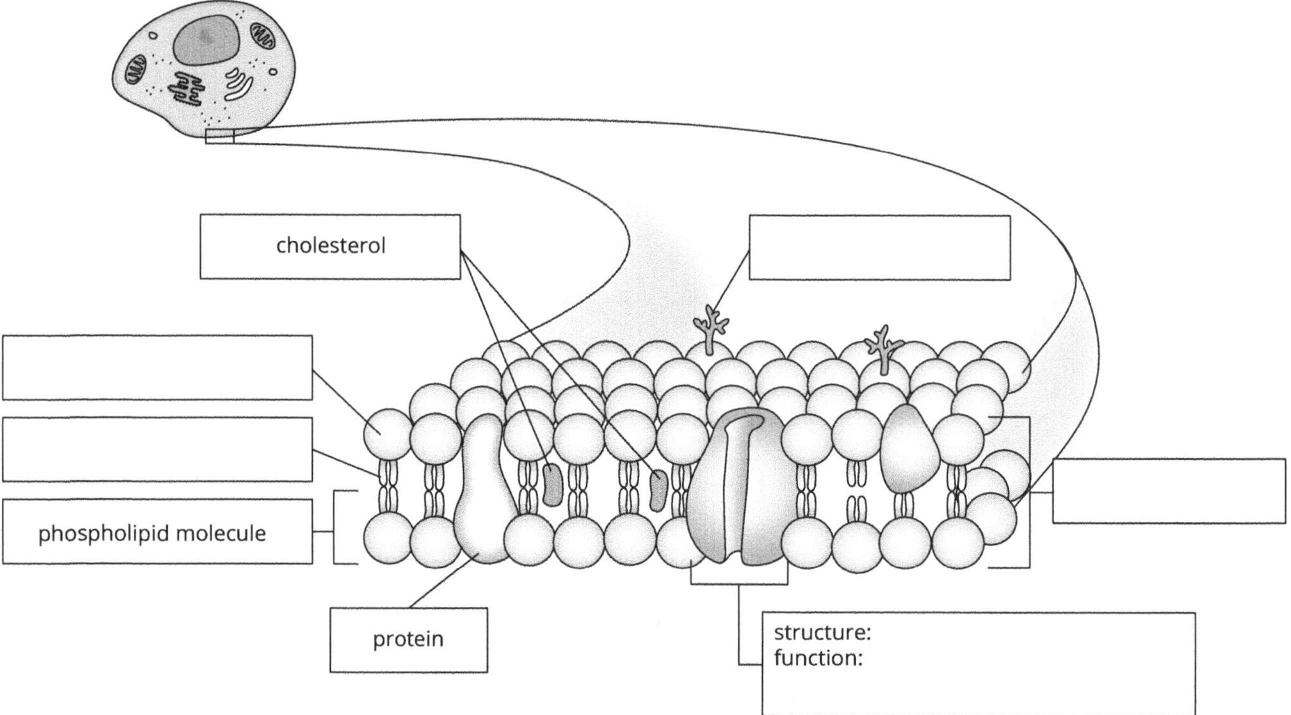

2 Outline the features of cell membranes that have led to their description as 'fluid-mosaic'.

3 Describe the role of the following molecules in cell membranes:

a cholesterol

b carbohydrate

4 Explain the link between the terms 'hydrophilic' and 'hydrophobic' in relation to the orientation of phospholipid molecules in cell membranes.

RATING MY LEARNING	My understanding improved	Not confident ◄—► Very confident ○ ○ ○ ○ ○	I answered questions without help	Not confident ◄—► Very confident ○ ○ ○ ○ ○	I corrected my errors without help	Not confident ◄—► Very confident ○ ○ ○ ○ ○

 ISBN 978 1 4886 1931 1

WORKSHEET 1.5

Selective cells—cell membranes and selectivity

Cell membranes act as the 'border guards' of cells, selectively determining which molecules enter and leave. As a result, cell membranes are described as selectively permeable or semipermeable.

The uptake of molecules by cells depends on several factors, including the concentration of those molecules on either side of the cell membrane. The route taken by different kinds of molecules as they enter and leave cells depends on the type of molecule and whether or not it is soluble in the phospholipid bilayer of the membrane.

Some very small molecules cross the cell membrane by passing between the lipid molecules within the membrane. Larger water-soluble molecules make their way through protein channels embedded across the phospholipid bilayer. Other molecules are actively taken up by cells through the protein channels against a concentration gradient. Yet others simply dissolve into the lipid of the cell membrane to enter cells. Refer to the figure in question 3 to answer each of the following questions.

1 Look carefully at the figure in question 3 and the different kinds of molecules represented, and their respective concentrations inside and outside of the cell. Decide on the process by which each enters the cell, giving reasons for your choice. Write your answers in the spaces provided.

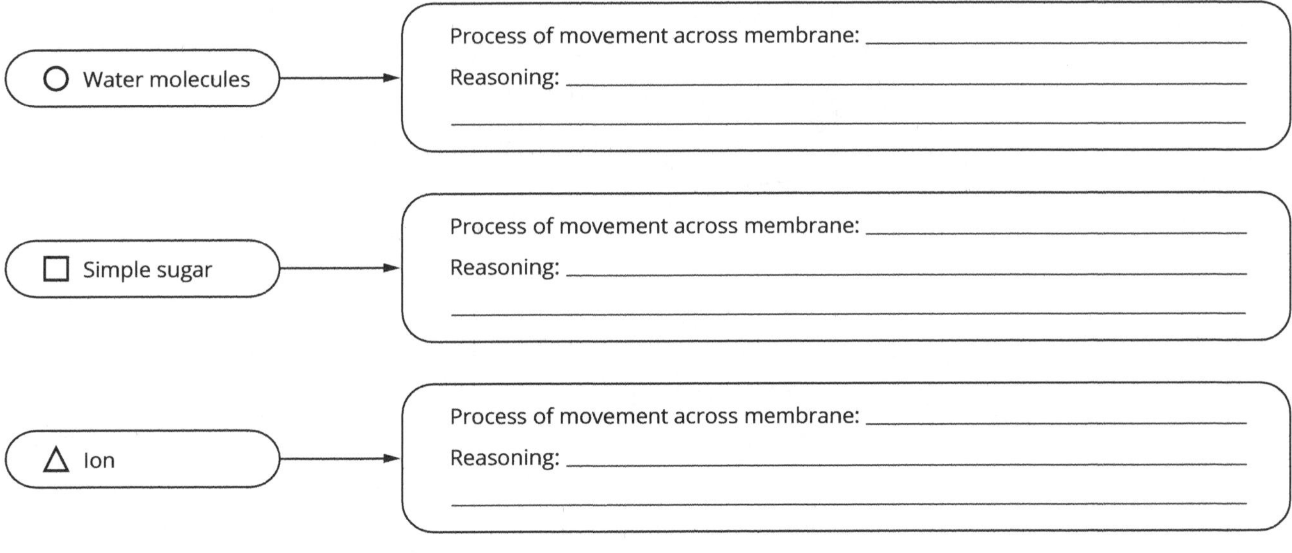

2 Molecule L enters the cell by dissolving into the lipid bilayer.

a What kind of molecules are represented by L?

b Name two kinds of molecules that enter or leave cells in this way.

ISBN 978 1 4886 1931 1

3 The cell membrane is composed of different kinds of molecules. Identify the various parts of the cell membrane indicated, providing information as directed.

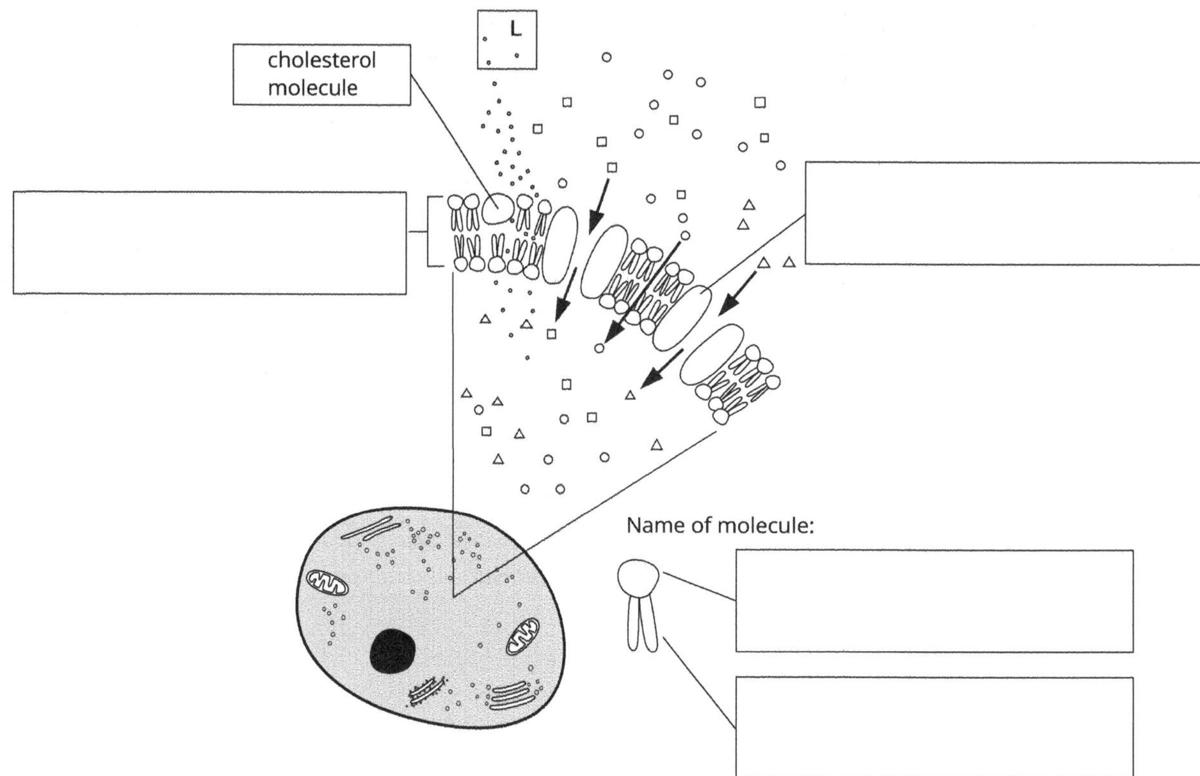

RATING MY LEARNING	My understanding improved	Not confident ◄——► Very confident ○ ○ ○ ○ ○	I answered questions without help	Not confident ◄——► Very confident ○ ○ ○ ○ ○	I corrected my errors without help	Not confident ◄——► Very confident ○ ○ ○ ○ ○

 ISBN 978 1 4886 1931 1

WORKSHEET 1.6

Reciprocal reactions—photosynthesis and cellular respiration

In photosynthesis, plant cells harness sunlight to drive a reaction in which carbon dioxide and water are combined to produce energy-rich organic molecules. These energy-rich organic molecules are, in turn, broken down to make energy available to meet the life-sustaining activities of cells.

1 Fill in the missing terms, definitions and equations in the spaces provided to construct a complete picture of the processes of photosynthesis and cellular respiration.

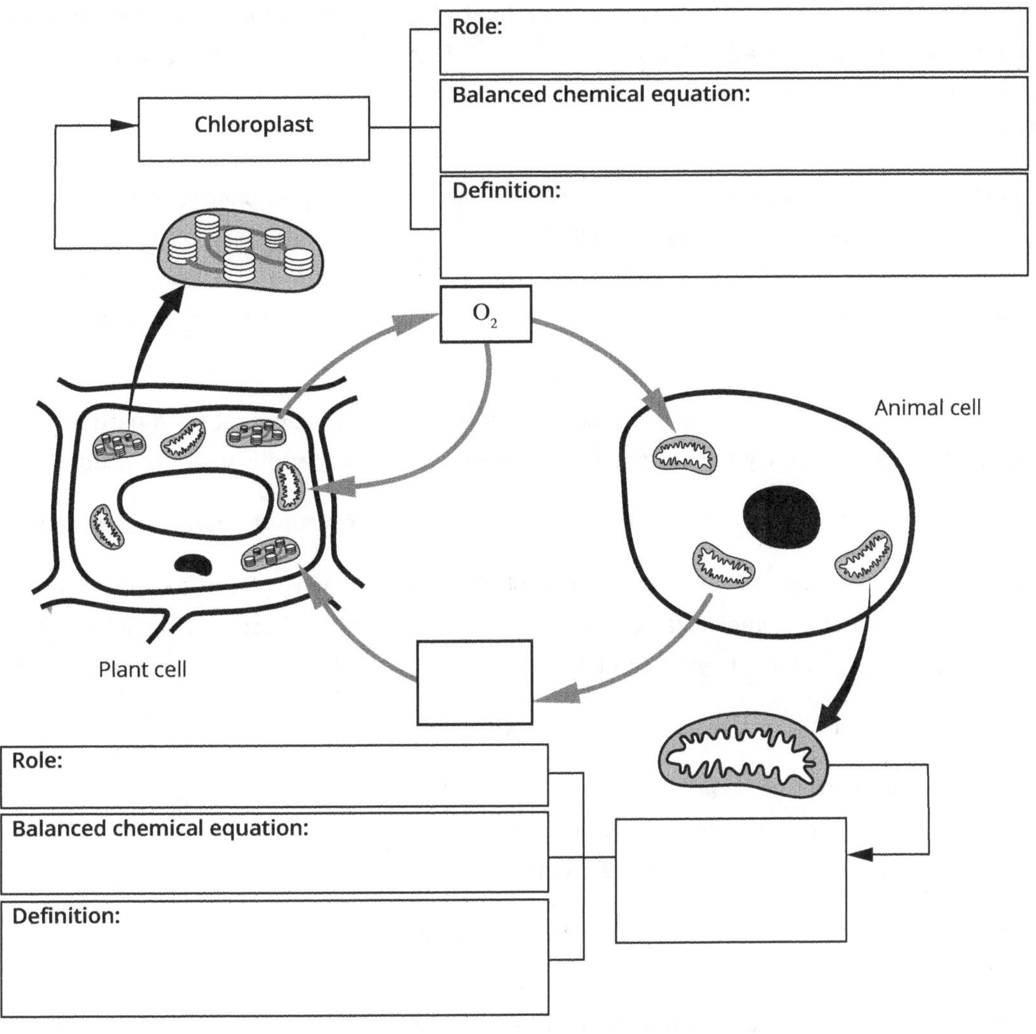

2 Summarise the relationship between the processes of photosynthesis and cellular respiration.

RATING MY LEARNING	My understanding improved	Not confident ◄—► Very confident ○ ○ ○ ○ ○	I answered questions without help	Not confident ◄—► Very confident ○ ○ ○ ○ ○	I corrected my errors without help	Not confident ◄—► Very confident ○ ○ ○ ○ ○

WORKSHEET 1.7

Enzyme ABC—amazing biological catalysts

1 Make a selection from the list below to fill in the missing words. This will give you a complete summary of enzymes and their functions.

denatured	deficiency	substrate-specific	temperature	protein
catabolic	catalysts	cellular respiration	metabolism	active site
photosynthesis	enzyme-substrate complex	irreversible	anabolic	coenzymes
regulatory	competitive	optimal		

Enzymes are biological ________________. They increase the rate at which chemical reactions occur in living cells. They are organic compounds composed of ________________. Enzymes are important facilitators of energy-transforming reactions, recycling processes, synthesis of some compounds and breakdown of others—all of the processes that make up the ________________ of cells.

Enzymes are involved in both the construction and breakdown of biological molecules. Chemical reactions in which larger molecules are constructed from smaller ones are called ________________ reactions. ________________ is an example of a reaction that typically occurs in plant cells. ________________ reactions involve the breakdown of large molecules into smaller ones. ________________ ________________ is an example of such a reaction in cells.

Enzymes are ________________ - ________________, that is, they have an active site that fits a particular substrate molecule only. This feature of enzymes is sometimes referred to as 'lock-and-key' or 'induced fit'.

Enzymes have ________________ conditions; that is, they operate most efficiently under particular conditions.

Enzymes are sensitive to factors such as ________________ and pH. Their rate of activity can also be influenced by the concentration of a substrate or enzyme. Cofactors and ________________ also affect enzyme action. The ________________ ________________ of an enzyme can be ________________ by excessive temperatures or pH.

Enzyme ________________ can result in disease.

________________ inhibitors are molecules that have a similar shape to the substrate. They compete for space in the active site of enzymes, preventing the formation of the ________________ - ________________ ________________, thereby reducing the rate of enzyme-catalysed reactions.

Enzyme inhibition can be reversible or ________________. Reversible inhibition has an important ________________ role in cells.

2 Circle the word(s) that make the following statements about enzymes correct.

a Enzymes are needed in small/large amounts.

b Enzymes are used up/not used up in chemical reactions.

c Enzymes can/cannot be used over and over again.

d Enzymes do/do not affect the final amount of product in a reaction.

e Enzymes increase/decrease the activation energy required to initiate reactions.

RATING MY LEARNING	My understanding improved	Not confident ◄——► Very confident ○ ○ ○ ○ ○	I answered questions without help	Not confident ◄——► Very confident ○ ○ ○ ○ ○	I corrected my errors without help	Not confident ◄——► Very confident ○ ○ ○ ○ ○

ISBN 978 1 4886 1931 1

WORKSHEET 1.8

Active enzymes—enzyme properties

Enzyme action

Enzymes catalyse chemical reactions in cells in two directions, that is, the reactions are reversible. The example below illustrates an enzyme catalysing the production of the disaccharide (two sugar) sucrose from the simple sugars glucose (G) and fructose (F).

1 Complete the diagram below by inserting the glucose and fructose molecules into the active site of the enzyme. Then draw the newly constructed molecule of sucrose after it is released from the enzyme.

Use coloured pencils to colour-code your diagram. Use different colours for each of the different kinds of sugar molecule and for the enzyme.

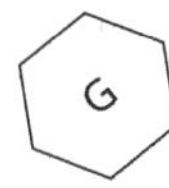

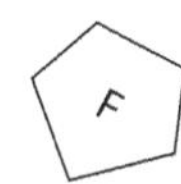

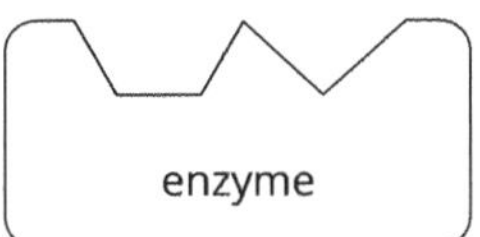

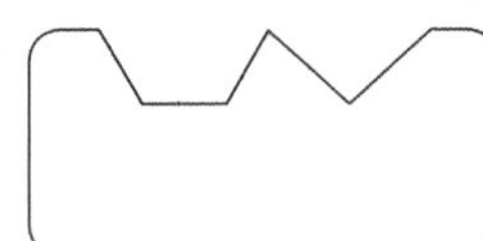

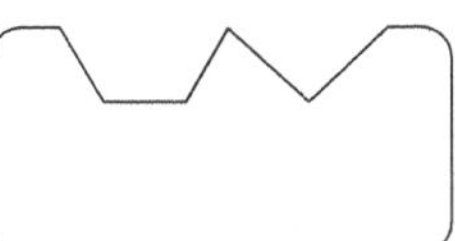

2 Explain why the relationship between enzymes and their substrates is described as 'lock-and-key'.

3 Name two factors that can affect the active site of an enzyme.

4 Describe what happens to the active site of an enzyme if it is denatured. What are the consequences for enzyme activity when this occurs?

5 Outline what is meant by the 'induced fit' model. Draw a diagram to illustrate.

RATING MY LEARNING	My understanding improved	Not confident ◄——► Very confident ○ ○ ○ ○ ○	I answered questions without help	Not confident ◄——► Very confident ○ ○ ○ ○ ○	I corrected my errors without help	Not confident ◄——► Very confident ○ ○ ○ ○ ○

ISBN 978 1 4886 1931 1

WORKSHEET 1.9

Literacy review—language to learn from

Fortunately, the language of science is designed to help us understand it. Some words may seem new and complex at first, but when we understand the origins of terms their use in various combinations actually makes sense.

In this activity you will consider the origins of words, and parts of words, to help make sense of their meaning. Scientific terms are often of Greek or Latin origin. Read the information about the terms set out below and use it as a guide to complete the missing meanings. Provide examples or further explanatory information where it is helpful to give extra context.

a	**Uni:** single **Multi:** many	**Meaning**
	Unicellular	Refers to an organism composed of a single cell.
	Multicellular	

b	**Pro:** early or primitive **Eu:** true **Karyon:** nut (nucleus)	**Meaning**
	Prokaryote	An organism characterised by the lack of a distinct, membrane-bound nucleus. Bacteria and Archaea are prokaryotes.
	Eukaryote	

c	**Intra:** inside **Extra:** outside	**Meaning**
	Intracellular fluid	Fluid that is present inside the cell.
	Extracellular fluid	

d	**Hydro:** water **Philic:** attracted to **Phobic:** repelled by	**Meaning**	
	Hydrophilic	Attracted to water; describes the phosphate head of the phospholid molecules that make up cell membranes of cells.	phosphate-containing head fatty acid tails
	Hydrophobic		

RATING MY LEARNING	My understanding improved	Not confident ◄——► Very confident ○ ○ ○ ○ ○	I answered questions without help	Not confident ◄——► Very confident ○ ○ ○ ○ ○	I corrected my errors without help	Not confident ◄——► Very confident ○ ○ ○ ○ ○

ISBN 978 1 4886 1931 1

WORKSHEET 1.10

Thinking about my learning

On completion of Module 1: Cells as the basis of life, you should be able to describe, explain and apply the relevant scientific ideas. You should be able to work with data, to interpret, analyse and evaluate it.

1 The following table lists the key knowledge covered in this module. Read each and reflect on how well you understand each concept. Rate your learning by shading the circle that corresponds to your level of understanding for each concept. It may be helpful to use colour as a visual representation. For example:

- green—very confident
- orange—in the middle
- red—starting to develop.

Concept focus	**Rate my learning** Starting to develop ◄——► Very confident				
Cell types—prokaryotes and eukaroytes, similarities and differences	○	○	○	○	○
Cell structure and ultrastructure, including roles of organelles	○	○	○	○	○
Cell membranes—structure and function	○	○	○	○	○
Movement of materials across membranes—diffusion, osmosis, facilitated diffusion, active transport	○	○	○	○	○
The cell theory, cell size and surface-area-to-volume ratio	○	○	○	○	○
Cell requirements—energy, matter and waste removal	○	○	○	○	○
Biochemical processes in cells—cellular respiration and photosynthesis	○	○	○	○	○
Enzyme activity in cells, features of enzymes, factors that affect enzyme activity	○	○	○	○	○

2 Consider points you have shaded from starting to develop to middle-level understanding. List specific ideas you can identify that were challenging.

____

____

____

3 Write down two different strategies that you will apply to help further your understanding of these ideas.

____

____

____

PRACTICAL ACTIVITY 1.1

Distinguishing cells—an observational activity

Suggested duration: 100 minutes

INTRODUCTION

This activity provides an opportunity to design and carry out a practical investigation into the similarities and differences between the cells of different kinds of organisms. You will need to be familiar with the use of the light microscope and accompanying equipment. You will also need to prepare some of your own slide specimens. If you are a little rusty on the procedures involved, your teacher may be able to arrange some refresher lessons for you before you begin this activity. See Investigating cells on pages 5–7 for details on using a light microscope and preparing specimens and biological drawings.

MATERIALS

- light microscope
- microscope slides
- coverslips
- onion
- *Spirogyra* (a filamentous green algae)
- iodine
- white tile or cutting board
- paper towel
- forceps
- scalpel
- selected prepared slides, for example:
  - protozoan
  - leaf epidermis
  - nerve cells
  - bacteria
  - white blood cells
  - cheek cells
  - cross-section of green plant stem
  - root hair tissue

PURPOSE

To design and carry out a laboratory investigation into the structural features of cells from different kinds of organisms.

To investigate the similarities and differences between cells from different kinds of organisms.

PROCEDURE

1 Use the materials listed to design a laboratory experiment that will allow you to investigate the structural features of cells from different kinds of organisms. You may wish to add further specimens to the suggested list—check this with your teacher.

2 Set out your experiment procedure in a numbered, step-by-step format. Write your instructions clearly so that another student could follow them easily. Your procedure should include instructions related to:

- mounting slides
- viewing fresh and prepared slides under the microscope
- preparing drawings of each specimen.

When you have completed the experimental design, have it checked by your teacher before proceeding with the laboratory work itself.

 Caution: safety instructions need to be provided in relation to the use of hazardous chemicals and sharp equipment.

Experimental design

ISBN 978 1 4886 1931 1

Cell drawings

3 Table 1 includes a number of features found in eukaryotic cells. Complete the table by indicating whether or not you were able to identify the features listed for each specimen you observed.

TABLE 1 Summary of cell features of organelles

Cell type	Features or organelles observed										Type of organism
	Cell membrane	Cell wall	Nucleus	Cytoplasm	Chloroplast	Mitochondria	Ribosomes	Vacuoles	ER	Golgi apparatus	

4 Decide upon a key to indicate that particular features were identified. For example, a tick may indicate that a feature has been clearly identified, a dash might be used to indicate that a feature was not observed.

Key

PROCESSING DATA

1 Suggest why it might not be helpful to use a cross to indicate that a feature has not been observed in a particular specimen.

2 Outline the role of the light microscope in the study of cells.

3 Outline any limitations that you encountered in this activity, and the impact they had on the investigation.

4 Suggest how these might be overcome in the future to ensure the greatest success in your experimental work.

CONCLUSION

5 Prepare a summary of your findings in this activity. Group the specimens according to common features. Include a discussion of:

- similarities between different kinds of cells
- differences between different kinds of cells.

RATING MY LEARNING	My understanding improved	Not confident ◀——▶ Very confident ○ ○ ○ ○ ○	I answered questions without help	Not confident ◀——▶ Very confident ○ ○ ○ ○ ○	I corrected my errors without help	Not confident ◀——▶ Very confident ○ ○ ○ ○ ○

 ISBN 978 1 4886 1931 1

PRACTICAL ACTIVITY 1.2

Shaping up—body shape versus diffusion

Suggested duration: 50 minutes

INTRODUCTION

In this activity blocks of agar jelly will simulate the shape of organisms. The jelly is a pink colour due to the presence of sodium hydroxide (an alkali) and phenolphthalein indicator. Phenolphthalein is colourless in acid. When the blocks are placed in an acid solution, the acid diffuses into the jelly, causing a colour change from pink to clear. The time taken for a block to totally decolourise is a measure of the rate of diffusion of acid into the jelly.

PURPOSE

To investigate the relationship between the surface-area-to-volume ratio and the diffusion rates of materials.

BACKGROUND

Sea lettuce (*Ulva lactuca*) is a green marine alga that, as its name suggests, looks like lettuce. It grows abundantly on sheltered rocky coasts. Its structure is very simple, as it consists of just two layers of cells. When the algae are submerged in water at high tide, materials enter the cells directly, and wastes leave the cells and diffuse into the surrounding water.

If a small piece of raw liver is put on the bottom of a freshwater pond, it will soon attract a number of small khaki-grey planarians, commonly called flatworms. Like *Ulva*, these worms are also very thin and flat. They have few internal organs. Gas exchange occurs directly across the body surface.

Both of these aquatic organisms depend on their shape to obtain nutrients and gases and remove wastes efficiently. Very few terrestrial organisms have this 'flattened' appearance (see figure at right).

PROCEDURE

1 Put on the disposable gloves, then accurately measure and cut a 2 cm × 2 cm block of jelly from each of the sheets provided (i.e. 2 cm, 1 cm and 0.5 cm thick sheets).

2 Half-fill a 250 mL beaker with sulfuric acid from the class stock.

3 Add the blocks of jelly to the acid. Watch until the first block becomes completely clear. In Table 1 write '1st' in red pen in the top of the column for the block that became clear first.

4 Using the spoon, immediately remove all of the blocks from the acid. Quickly pat the blocks dry with paper towel. Cut them in half with the scalpel and measure, in mm, the depth of the clear layer in each block. Record the depth of the clear layer in each block, that is, the depth to which the acid has penetrated each block.

5 Dispose of the acid and the jelly blocks as directed by your teacher.

PROCESSING DATA

1 a Calculate the surface area for each block of jelly and enter the data into Table 1.

b Calculate the volume of each block and enter the data into Table 1.

c Use a calculator to find the surface-area-to-volume ratio for each of the blocks. Enter the data into Table 1.

MATERIALS

- prepared sheets of agar–phenolphthalein jelly
  - 2 cm thick
  - 1 cm thick
  - 0.5 cm thick
- 0.1 mol L^{-1} sulfuric acid (class supply)
- 250 mL beaker
- plastic teaspoon
- scalpel
- clear plastic ruler
- tile or large Petri dish lid
- strip of paper towel
- disposable gloves
- calculator

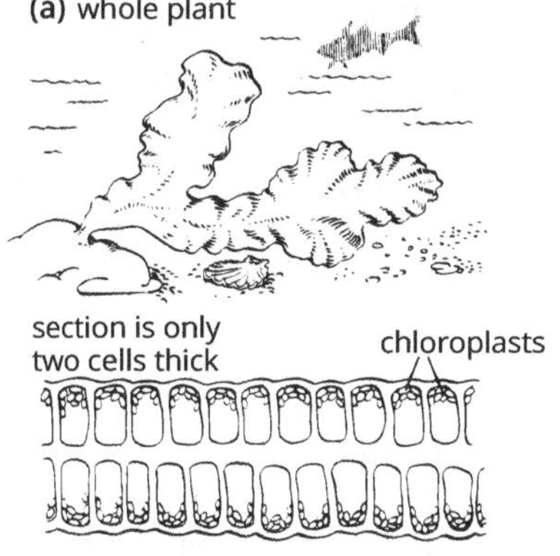

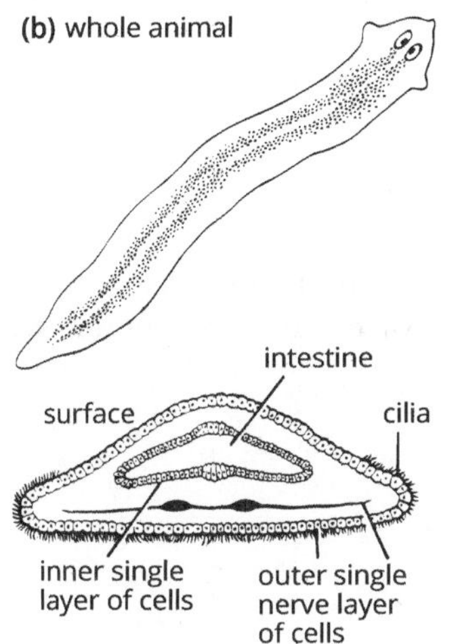

(a) Sea lettuce (Ulva lactuca) and (b) a planarian, commonly called a flatworm

TABLE 1 Jelly measurements

Block number	**1**	**2**	**3**
Block dimensions (cm)	2 × 2 × 2	2 × 2 × 1	2 × 2 × 0.5
Depth of clear layer of block (mm)			
Surface area (SA, cm^2)			
Volume (V, cm^3)			
Surface-area-to-volume ratio (SA : V)	: 1	: 1	: 1

DISCUSSION

2 Explain what has caused the blocks to become clear. Include a discussion of both substances and processes.

3 a Which block was the first to become completely clear?

b From your observations of the other two blocks, predict which would be the next to become clear. Explain why you think so.

4 a Which block has the largest surface area (SA)?

b Which block has the largest volume (V)?

c Which block has the largest surface area in proportion to its volume (i.e. surface-area-to-volume ratio or SA : V)?

5 Describe what happens to the SA : V of a block as its size gets smaller.

CONCLUSION

6 Describe the relationship between the size and shape of a block and the rate at which acid diffuses into it.

7 Given that cells need to obtain all the materials required for life from the surrounding environment and remove wastes to the surrounding environment, suggest why cells are usually microscopic in size.

RATING MY LEARNING	My understanding improved	Not confident ◄——► Very confident ○ ○ ○ ○ ○	I answered questions without help	Not confident ◄——► Very confident ○ ○ ○ ○ ○	I corrected my errors without help	Not confident ◄——► Very confident ○ ○ ○ ○ ○

ISBN 978 1 4886 1931 1

PRACTICAL ACTIVITY 1.3

Partially permeable membranes—diffusion and osmosis

Suggested duration: Part A—50 minutes; Part B—50 minutes

PURPOSE

To investigate the movement of particles in solution across a semipermeable membrane.

MATERIALS

- dialysis tubing
- iodine/potassium iodide
- Clinistix™
- 5% starch solution
- glucose solution
- thistle funnel
- 2 x gas jars
- retort stand and clamp
- test-tubes
- test-tube rack
- rubber bands
- 50 mL beaker

BACKGROUND

To test for the presence of starch, add a few drops of iodine/potassium iodide solution to the test solution. If starch is present the solution will change to a blue-black colour.

To test for the presence of glucose, dip a Clinistix into the test solution. If glucose is present, the Clinistix will change from pink to purple.

Part A—presenting partial permeability

PROCEDURE

1 Pour glucose solution into a 50 mL beaker to a depth of about 1 cm. Dip in the coloured tab of a Clinistix. Describe your observations in Analysis of results 1.

2 Fill a gas jar with water until it is about three-quarters full. Test the water in the same way with a new Clinistix. Note your result in Analysis of results 2.

3 Set up the equipment as shown in the figure at right.

Set up the retort stand to support the thistle funnel first. Then secure the dialysis tubing. Moistening the dialysis tubing will help to open the ends. Tie one end firmly closed with a rubber band. Tie the other end so that it is securely fastened to the thistle funnel. Gently pour glucose solution into the thistle funnel until the level rises about 2 cm up the stem of the funnel. Place the gas jar beneath the thistle funnel. Use the clamp to lower the dialysis tubing completely into the water solution.

4 Leave the set-up undisturbed for 30 minutes (longer if possible). Use the Clinistix to once again test the solution in the gas jar. Describe your observations in Analysis of results 3.

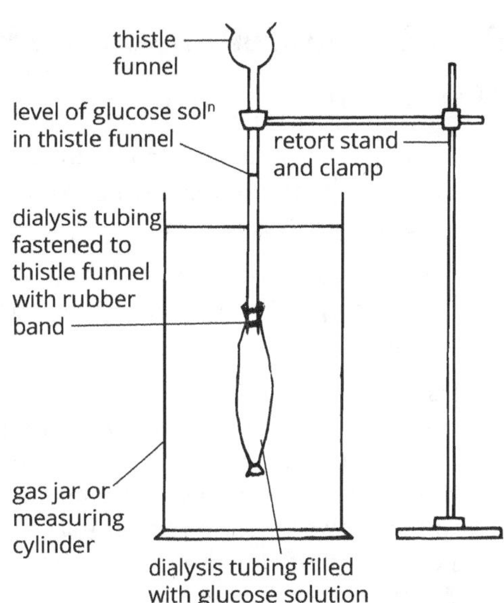

ISBN 978 1 4886 1931 1

ANALYSIS OF RESULTS

1 Describe your observations after step 1 of the procedure.

2 a Note the result from step 2 of the procedure. Describe the Clinistix colour for water in the gas jar.

b Explain why you have tested the water in the gas jar with the Clinistix.

3 a Describe your observations after step 3 of the procedure.

b Explain what has happened.

Part B—modelling osmosis

PROCEDURE

1 Pour starch solution into a test-tube to a depth of about 1 cm. Add a few drops of iodine/potassium iodide solution. Describe any colour change that occurs in Analysis of results 1.

2 Fill the second gas jar with water until it is about three-quarters full. Add several drops of iodine/potassium iodide solution. Answer the questions in Analysis of results 2.

3 Use the equipment listed to construct an experimental set-up similar to the one shown in the figure at right. Prepare the set-up in the same way you did for Part A, except that this time you will pour starch solution into the thistle funnel/dialysis tubing. Place the gas jar beneath the thistle funnel. Lower the dialysis tubing completely into the water and iodine/potassium iodide solution. Leave the set-up undisturbed for at least 30 minutes.

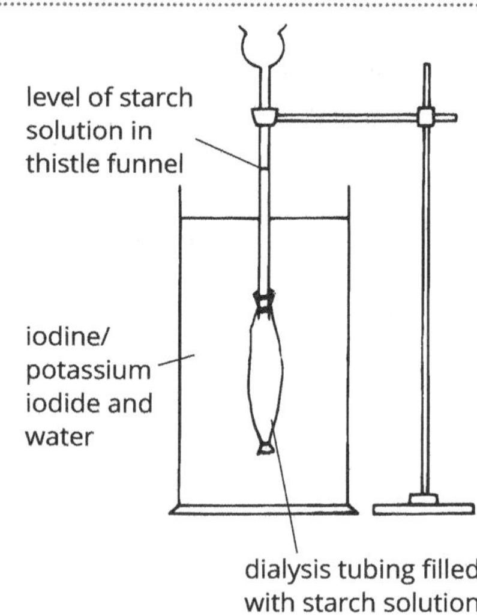

ANALYSIS OF RESULTS

1 Describe any colour change that occurs when iodine solution is added to the starch.

2 a What colour is the solution in the gas jar?

b What does the colour of the solution indicate about the presence of starch?

 ISBN 978 1 4886 1931 1

3 Use coloured pencils to colour and label the figures to reflect your observations at the start of the activity and again after 30 minutes.

start of activity

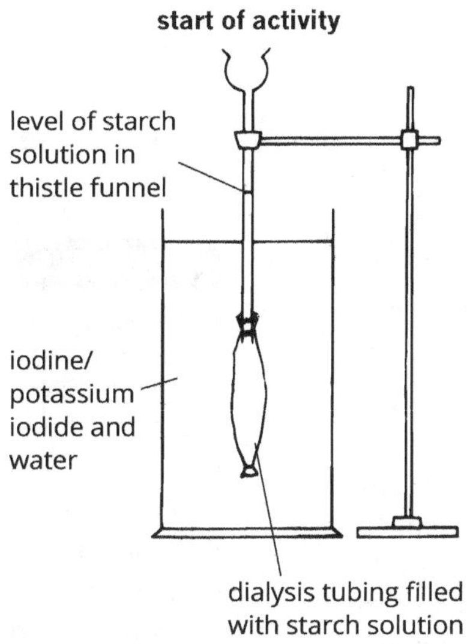

after 30 minutes

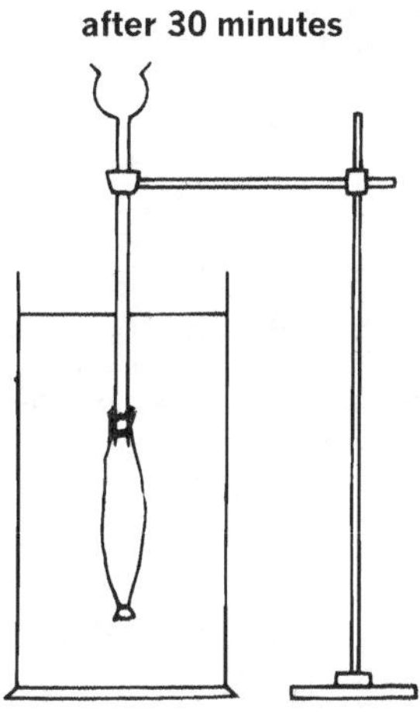

4 Describe any changes you observe.

5 Account for the colour changes you have observed. Where did the molecules of water, iodine/potassium iodide and starch begin and where did they move to?

CONCLUSION

In this activity, you considered evidence that indicates some kinds of molecules move across semipermeable membranes while others do not. Write the definitions for:

1 **a** osmosis

b diffusion

2 Explain why some kinds of molecules are able to pass through semipermeable membranes and others are not. Use specific examples.

RATING MY LEARNING	My understanding improved	Not confident ↔ Very confident ○ ○ ○ ○ ○	I answered questions without help	Not confident ↔ Very confident ○ ○ ○ ○ ○	I corrected my errors without help	Not confident ↔ Very confident ○ ○ ○ ○ ○

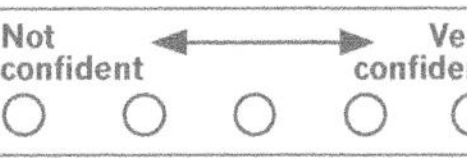
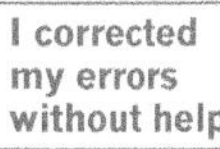

PRACTICAL ACTIVITY 1.4

Paramount pathways—photosynthesis and cellular respiration

Suggested duration: 60 minutes

PURPOSE

To investigate factors that affect the process of photosynthesis in an aquatic plant.

To consider the relationship between the processes of photosynthesis and cellular respiration.

BACKGROUND

Photosynthesis and cellular respiration are considered reciprocal reactions in the biochemical pathways of plants—the products of one reaction become the raw materials for the other. In photosynthesis, green plants combine carbon dioxide and water in the presence of light and chlorophyll, producing glucose and oxygen. During cellular respiration, glucose and oxygen are consumed to release energy for cells and carbon dioxide and water are produced.

When carbon dioxide is dissolved in water a chemical reaction takes place that results in the production of carbonic acid. The reaction can be summarised as follows:

carbon dioxide + water → carbonic acid

$$CO_2 + H_2O \rightarrow H_2CO_3$$

A simple test for the presence of dissolved carbon dioxide in water involves the indicator bromothymol blue. Bromothymol blue is sensitive to changes in pH, turning blue in alkaline conditions and yellow in acidic conditions.

Elodea is a common pond weed that uses carbon dioxide in photosynthesis and generates carbon dioxide in cellular respiration. As *Elodea* is an aquatic plant, gaseous exchange occurs directly with its watery environment. In this experiment you will test for the presence of carbon dioxide in different conditions.

MATERIALS

- *Elodea* (pond weed) or *Spirogyra* (a filamentous green algae)
- 2 x test-tube racks
- 4 x test-tubes
- 4 x rubber stoppers
- pond water
- marker pen
- bromothymol blue indicator solution
- spotting tile
- distilled water
- ammonia
- dilute hydrochloric acid
- 24-hour light source

Elodea is a known noxious weed and is banned in some states and territories. Safe disposal is recommended.

PROCEDURE

Pre-test

1. Add a couple of drops of dilute hydrochloric acid to one well on your spotting tile.
2. Add a drop of bromothymol blue and record the colour in Table 1.
3. Repeat steps 1 and 2 using distilled water and ammonia.

TABLE 1 Colour change with bromothymol blue

Solution	Colour
dilute hydrochloric acid	
distilled water	
ammonia	

 ISBN 978 1 4886 1931 1

Experiment

4 Set up two test-tube racks, each with two test-tubes. Use the marker pen to clearly label the test-tubes A, B, C and D, as shown in the figure below.

5 Fill each of the test-tubes with pond water until they are almost full. Add 20 drops of bromothymol blue to each of the four test-tubes.

6 Place a long piece of *Elodea* into each of test-tubes B and D. Make sure the two pieces are of approximately equal length.

7 Tightly stopper each of the test-tubes.

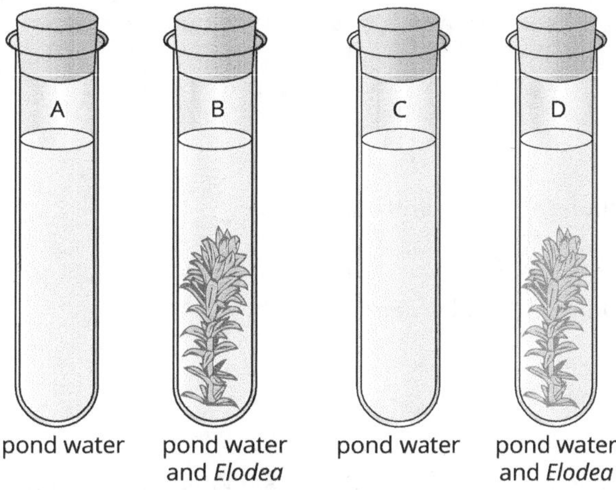

8 Record the colour of the solution in each test-tube in Table 2.

9 Place test-tubes A and B under a 24-hour light source and test-tubes C and D in the dark.

10 Leave the experiment set up for a minimum of 24 hours.

TABLE 2 Results for indicator colour

	Test-tubes in the light		Test-tubes in the dark	
	A	B	C	D
Initial colour				
Final colour				

Next lesson (at least a full day later)

11 Check each of the test-tubes and record the final colour of the solution for each in Table 2.

DISCUSSION

1 Write a hypothesis for this investigation.

2 **a** Which test-tubes represent the control(s) in this experiment?

b Outline the purpose of the control.

PRACTICAL ACTIVITY 1.4

3 Describe two ways in which you controlled this experiment.

4 What does a change in colour suggest about the pH of the solution?

5 Which test-tube(s) in this experiment showed a change in colour? Explain why this occurred.

6 Write out the balanced equation for photosynthesis.

7 Write out the balanced equation for cellular respiration.

CONCLUSION

8 a State one conclusion you can draw from the experiment set up in the light.

b Outline the evidence that supports this conclusion.

9 a State one conclusion you can draw from the experiment set up in the dark.

b Outline the evidence that supports this conclusion.

10 Summarise the relationship between the biochemical pathways of photosynthesis and cellular respiration in *Elodea*.

RATING MY LEARNING	My understanding improved	Not confident ◄——► Very confident ○ ○ ○ ○ ○	I answered questions without help	Not confident ◄——► Very confident ○ ○ ○ ○ ○	I corrected my errors without help	Not confident ◄——► Very confident ○ ○ ○ ○ ○

 ISBN 978 1 4886 1931 1

PRACTICAL ACTIVITY 1.5

Capable catalase—investigating enzyme efficiency

Suggested duration: 60 minutes

INTRODUCTION

The metabolism of cells involves many different chemical reactions. These include energy-releasing reactions such as cellular respiration, as well as others in which substances are constructed or digested. The byproducts of some of these chemical reactions are harmful and must be removed from the body before they accumulate to a level at which they can cause tissue damage.

Carbon dioxide is one example of a metabolic waste produced by cells. It is carried to the lungs via the bloodstream, where it is removed from the body during exhalation. Hydrogen peroxide is another waste product of cell metabolism. It is a potentially harmful chemical and must be removed immediately. The enzyme catalase operates in cells to continually break down hydrogen peroxide into harmless products.

MATERIALS

- liver
- chopping board
- newspaper
- scalpel
- test-tubes
- test-tube rack
- hydrogen peroxide
- Bunsen burner
- heat mat
- tripod
- gauze mat
- beaker of water
- mortar and pestle
- splint
- matches
- protective eye glasses
- disposable gloves
- stopwatch

PURPOSE

To design and conduct an investigation of enzyme activity.

To consider factors that affect enzyme activity.

PROCEDURE

Use the list of materials provided to design an experimental procedure to test whether or not:

- liver cells contain the enzyme catalase
- temperature affects enzyme activity.

Use the experimental set-up below right as a guide.

When you have completed your experimental design, check with your teacher before conducting your experiment.

 Caution: care is required when using sharp equipment such as a scalpel, and chemicals such as hydrogen peroxide. Check any special procedures with your teacher.

Set-up of an experiment for investigating enzyme efficiency

ISBN 978 1 4886 1931 1

ANALYSIS OF RESULTS

1 Describe any evidence you observed that enzyme activity was taking place.

2 a How long did the activity you observed in the test-tube last?

b What could you do to make the activity in the test-tube continue?

c What does this suggest about the way in which enzymes are involved in the chemical reaction?

3 The chemical equation below describes part of the reaction taking place.

$$2H_2O_2 \xrightarrow{\text{catalase}} 2H_2O + \underline{\qquad\qquad}$$

a Look carefully at the input into this reaction. Suggest what gas is being produced.

b Suggest how you could test for the kind of gas produced. (Clue: look at the materials listed for this activity that you haven't used yet.)

4 Describe any differences you observe in the rate of enzyme activity for a small block of liver compared with the same amount of liver that has been processed using the mortar and pestle. Account for the differences you observe.

ISBN 978 1 4886 1931 1

PRACTICAL ACTIVITY 1.5

5 a Describe your observations when liver (block and processed) that has been boiled is exposed to hydrogen peroxide.

b Explain your observations.

CONCLUSION

6 Describe evidence from this investigation that supports the hypothesis that 'enzymes are not used up in the chemical reactions they catalyse and can be reused'.

7 Using your understanding of enzymes and referring to the experimental results achieved in this investigation, outline the effect of temperature on enzyme activity.

8 Referring to your experimental results, describe one other factor that affects enzyme activity.

RATING MY LEARNING	My understanding improved	Not confident ◄──► Very confident ○ ○ ○ ○ ○	I answered questions without help	Not confident ◄──► Very confident ○ ○ ○ ○ ○	I corrected my errors without help	Not confident ◄──► Very confident ○ ○ ○ ○ ○

DEPTH STUDY 1.1

Enterprising enzymes—investigating enzyme activity

Suggested duration: Part A—90 minutes; Part B—120 minutes

INTRODUCTION

In this activity you will consider the importance of enzymes in living things by investigating a plant enzyme (primary investigation) and an enzyme in humans (secondary-sourced investigation).

In Part A you will plan and conduct an investigation that further develops your knowledge and understanding of enzyme activity. The investigation involves developing a question and hypothesis, which you will test. You will make predictions based on your knowledge and understanding of enzymes, then conduct the experiment you have designed. After recording the experimental data you will interpret and analyse it. You will communicate the findings of your investigation by completing the practical report, following the standard structure of a scientific report. To guide you through this a template is provided below.

As your experience in conducting laboratory investigations increases, so does your familiarity with the elements of a sound scientific method. You will be guided through the planning and conducting of this activity with a view to working more independently in subsequent investigations.

In Part B you will focus on an enzyme that is absent in humans suffering a selected enzyme deficiency disease. This will be a secondary-sourced investigation in which you will research the enzyme and its role, and communicate your findings in a digital presentation.

MATERIALS

- 3 x water baths set at 0°C, 20°C and 40°C
- 3 x thermometers
- 3 x clean test-tubes (approximately 20 mL)
- test-tube rack
- 2 x 10 mL measuring cylinders
- iodine solution
- 1% starch solution
- 1% diastase solution
- marker pen
- spotting tile
- 2 x dropping pipettes

PURPOSE

To investigate the effect of temperature on the activity of the plant enzyme diastase.

To investigate the role of a selected enzyme in the human body.

Part A—plant enzyme diastase

BACKGROUND

The cells of living organisms are like tiny factories, each the centre of a great deal of activity, including biochemical reactions that involve the breakdown of some molecules and the construction of others. Enzymes are the biological catalysts that facilitate these biochemical reactions. Without them, critical chemical reactions cannot occur or occur too slowly to maintain the wellbeing of cells. As a result, cell function is compromised, sometimes leading to cell death and the death of the organism.

As well as cellular respiration and photosynthesis, many other life-sustaining biochemical pathways occur in cells. In plant cells, the carbohydrate-reducing enzyme diastase (an amylase) is an important enzyme involved in converting starch in storage organs to simple sugars that can be transported to growing tissue.

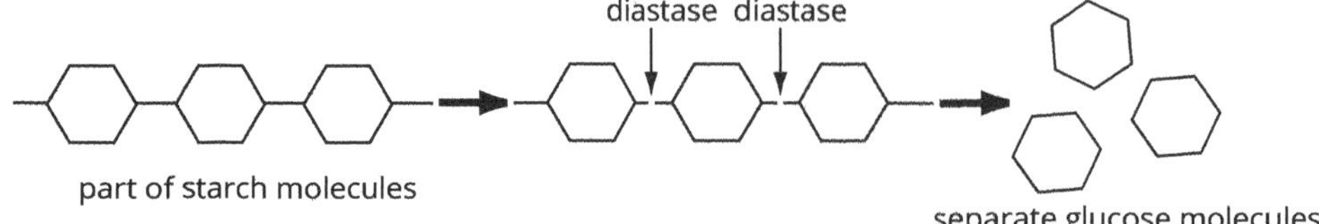

Effect of diastase on starch

Starch is a complex carbohydrate molecule made up of many glucose molecules (single sugars) joined together by chemical bonds. Starch stains a deep blue-black colour when mixed with iodine solution. Glucose and other simple sugars are not affected by iodine solution and so remain the yellow/brown colour of the iodine solution.

 ISBN 978 1 4886 1931 1

QUESTIONING AND PREDICTING

1 Laboratory investigations inevitably commence with observations and questions. In view of the purpose of this investigation, there are several questions we could ask. For example:

- Does temperature have an effect on the activity of enzymes?
- Do extremes of temperature have the same effect on enzymes?

Write out two more questions raised by the consideration of enzymes and temperature.

2 Refer to the purpose of this investigation and the two questions you have posed. Use these to develop a hypothesis that can be tested in this investigation. Write your hypothesis in the space provided below.

PLANNING YOUR INVESTIGATION

3 Consider the list of materials as the starting point for this activity. How could you use these materials to set up an experiment to test your hypothesis? Use the space below to write out a logical sequence of steps as the procedure for this investigation. Remember to use detailed, numbered, step-by-step instructions that another student could easily follow. Procedural steps should be written in the third person. It will be important to check your procedure with your teacher before commencing the activity. This includes safety considerations and disposal steps.

4 You will also need to think about effective ways of recording your experimental data. Experimental data is recorded under 'Results' in your report. Tables are an effective way of recording data. Table 1 has already been prepared for you. Think about why you would need to record this information. You will need a separate table for the experimental data.

5 As well as keeping a record of your experimental results, scientific reports record other important information about the investigation under the heading 'Discussion'. Often you will do this by answering set questions that guide the structure of your discussion. When you plan the investigation, you need to ensure these items are addressed. They include:

- identifying the variables
- describing ways in which the experiment was controlled
- interpreting the experimental data—summarising what the data means (Processing data and information)
- analysing the experimental data—explaining what the data tells you in terms of the purpose and hypothesis (Analysing data and information)
- evaluating the scientific method—this includes identifying any limitations and how these might have impacted on the investigation, and suggestions about how to reduce or eliminate these in future investigations
- describing and discussing other relevant observations or information.

6 The final element of a scientific report is the conclusion. This is a clear and concise statement that summarises the findings of the investigation. The conclusion should respond to the purpose and state whether the hypothesis is supported or not supported, drawing on experimental evidence to support this. Write your conclusion for this investigation in the space provided on the next page.

CONDUCTING YOUR INVESTIGATION

Procedure

PROCESSING DATA AND INFORMATION

Results

TABLE 1 Colour results

Solution	Iodine
starch	
diastase	

COMMUNICATING

Discussion

CONCLUSION

Part B—a selected human enzyme

PLANNING YOUR INVESTIGATION

Enzymes play critical roles in the ongoing metabolism in the cells of all organisms. In Part B you will select a human enzyme and investigate its role and the consequences of insufficient supply or absence of the enzyme. Such conditions are called enzyme-deficiency diseases. It is only when these enzymes are in short supply or missing that we are fully able to appreciate the critical nature of their presence. Such diseases have long been the subject of scientific and medical research. This research informs us about the role of specific enzymes and their importance.

 ISBN 978 1 4886 1931 1

In this secondary-sourced investigation, your research will help to answer questions but is sure to raise others. Keep a record of questions that arise from your research and include them in your presentation. You are not required to answer the questions that might arise, but they demonstrate a deeper level of thinking and consideration, and reflect society's need for information. Such questions drive further research that assists scientists in getting closer to solutions that improve the quality of life for people with these conditions.

Some examples of enzyme-deficiency diseases are listed below. You may select one of these or choose another in consultation with your teacher.

- Lactose intolerance
- Galactosaemia
- Globoid cell leukodystrophy (Krabbes disease)
- Phenylketonuria (PKU)
- Tay-Sachs disease

COMMUNICATING

Criteria

You are required to address each of the following criteria in your investigation:

a Name the deficiency disease and identify the enzyme missing in sufferers of this condition.

b Outline the role of this enzyme in human metabolism.

c Describe the symptoms of the enzyme deficiency disease.

d Describe treatments/management strategies.

e Outline the prognosis for patients given current treatments/management.

f Present statistical data about the prevalence of this condition.

g Name and describe medical technologies/drugs that are currently the subject of research and development of more effective treatments and potential cures.

h Include other relevant information around the condition, for example, approximate age when symptoms typically develop, cause of condition (genetic or other triggers).

i Record any questions that arise from the research.

j Summarise the significance of enzymes in the functioning of cells and the body.

Guidelines

i You will communicate your findings about your selected enzyme and the related enzyme-deficiency disease using a digital format of your choice. This is an opportunity to showcase your research and communication skills. There are some important protocols that you are required to follow. These include:

- writing your own notes from your research. Writing these in your own words demonstrates your understanding of the information
- using your research notes to plan and construct your presentation of the information
- using appropriate terminology to demonstrate you are familiar with the language of the subject and the topic
- defining new terms for your audience, again, in your own words
- using footnotes to reference factual information, scientific studies and data
- preparing a bibliography according to expected guidelines.

ii It will be helpful to refer to page xiv of the Biology toolkit to guide you through the elements of a secondary-sourced investigation. You will be required to locate relevant and reliable resources, organise your notes, analyse the information you uncover and organise and present it in an appropriate fashion. GO TO ➤ page xiv

iii As a guide to length for this task, a student selecting a PowerPoint presentation for delivery should employ a maximum of 12 slides, including the title slide and bibliography.

iv Keep your work concise and to the point. You will be assessed on addressing each criterion logically and clearly, not on volume.

Multiple choice

1 Examine the electron micrograph of a eukaryotic cell.

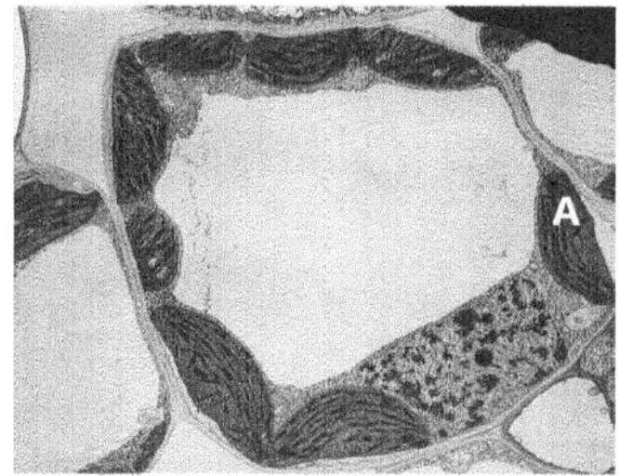

The organelle labelled A in the cell above is a:

A vacuole

B Golgi apparatus

C chloroplast

D mitochondrion

2 The ratio between surface area and volume is critical to cell survival as it determines how quickly materials can be exchanged between the internal and external environments. Examine cells A–D below. All lengths are in µm. The surface area of each cell is shown beneath it. All the cells have a volume of 308 µm³.

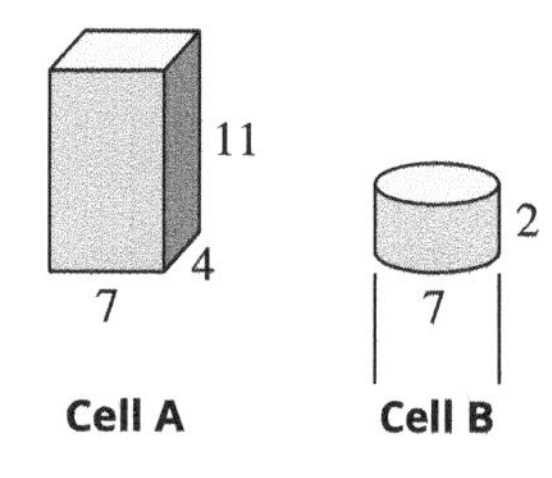

surface area: 298 μm^2 241.9 μm^2

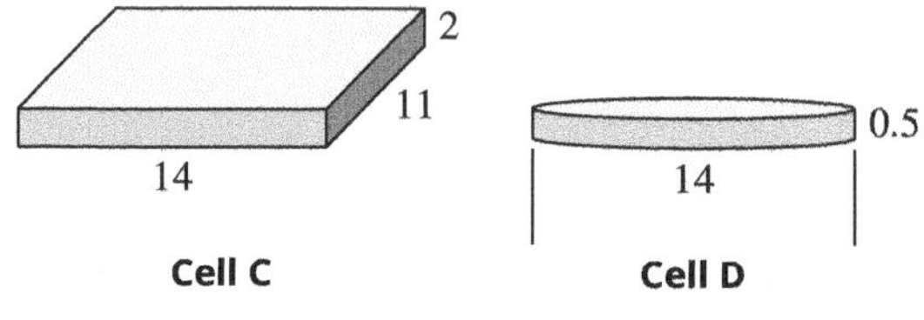

surface area: 408 μm^2 633.35 μm^2

The cell that will exchange materials most rapidly with its environment is:

A Cell A

B Cell B

C Cell C

D Cell D

The following diagram is relevant to questions 3 and 4. It shows molecules crossing the cell membrane.

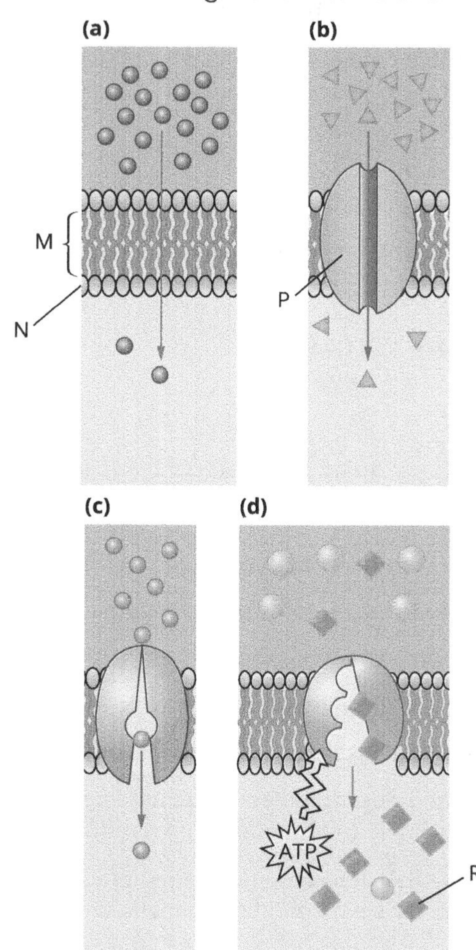

3 Which statement correctly identifies molecules M, N, P and R?

A M is phosphate, N is fatty acid, P is protein and R is glucose

B M is fatty acid, N is phosphate, P is protein and R is glucose

C M is phosphate, N is protein, P is glucose and R is fatty acid

D M is fatty acid, N is phosphorus, P is protein and R is glucose

4 Active transport is represented by:

A diagram (a)

B diagram (b)

C diagram (c)

D diagram (d)

 ISBN 978 1 4886 1931 1

5 The graph below shows the rate of photosynthesis in two plants as light intensity increases.

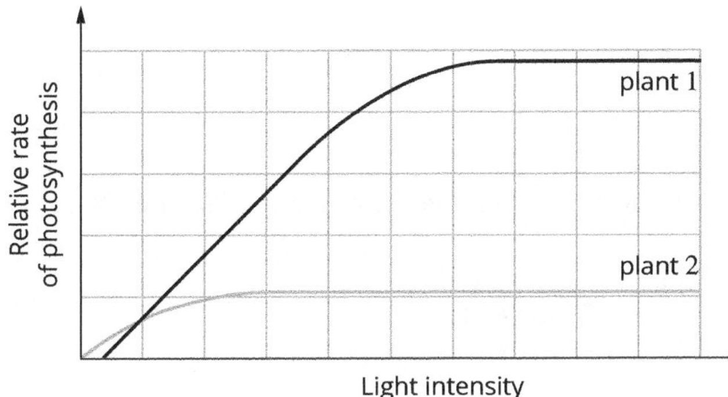

The factor most likely to cause the difference in the rate of photosynthesis between plant 1 and plant 2 is:

A water availability

B oxygen availability

C carbon dioxide availability

D light intensity

6 Which statement in relation to enzyme activity in cells is not true?

A Enzymes are inorganic catalysts that facilitate chemical reactions in cells.

B Enzymes are not consumed in the chemical reactions they facilitate, and so are available to be used again and again.

C Enzymes may be denatured when exposed to excessively high temperatures and extremes of pH.

D Enzymes reduce the activation energy of chemical reactions.

Short answer

7 Examine the animal and plant cells shown in solutions of different concentrations in the following table.

a In each instance, identify the solution in the beaker as isotonic, hypertonic or hypotonic. Enter your responses in the table below, with an explanation for each.

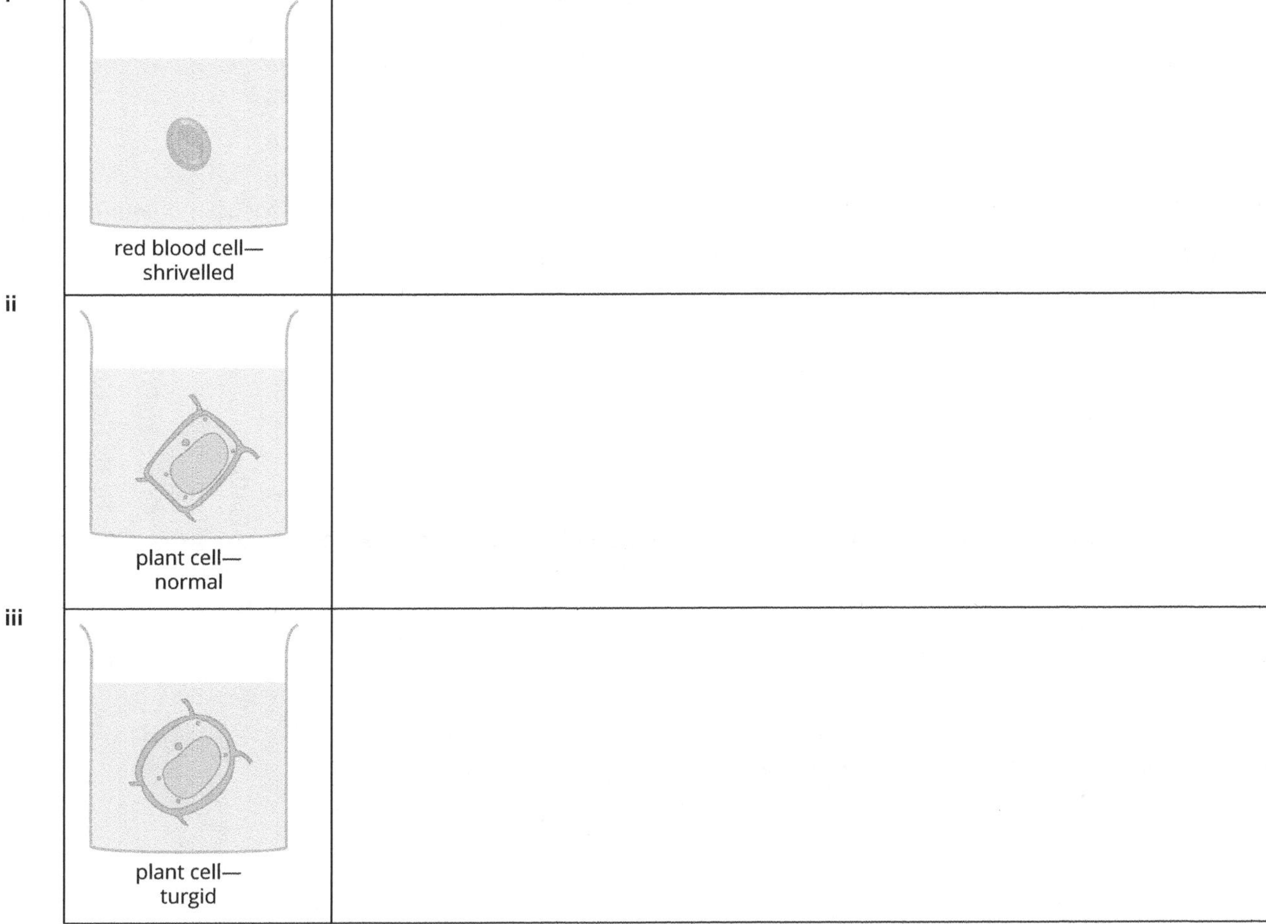

b Suggest why the plant cell shown in (iii) does not burst, despite the uptake of fluid.

8 Consider diagrams (i) and (ii) below, representing different models of how enzymes might interact with their substrate molecules.

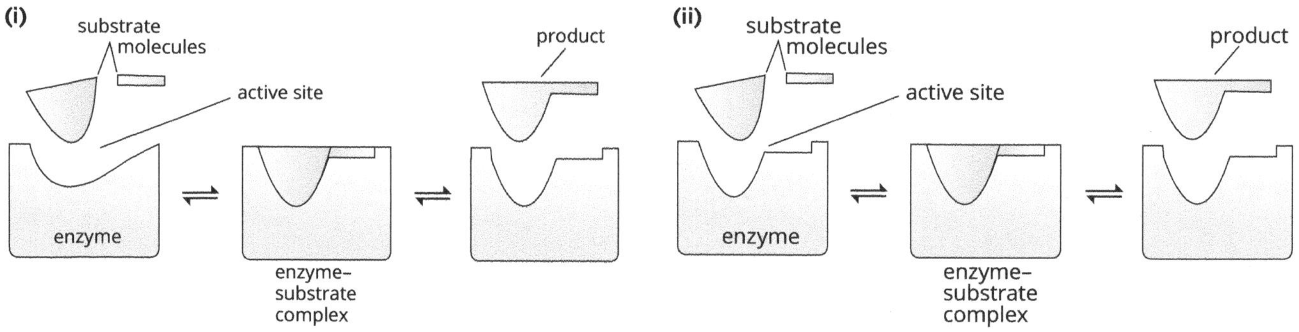

a What is meant by the term 'specificity' as it relates to enzymes? Give an example to illustrate your understanding.

b Use the diagrams to distinguish between the 'lock-and-key' model and the 'induced fit' model of enzyme-substrate interaction. Provide an explanation of the differences between the two.

9 Photosynthesis and cellular respiration are examples of biochemical processes that occur in different kinds of cells.

a Write out the balanced chemical equation for both processes.

Photosynthesis:

Cellular respiration:

b Suggest why these two biochemical reactions are described as inverse reactions.

c Outline the significance of these two processes for living things.

 ISBN 978 1 4886 1931 1

Extended response

10 A biology class studying enzyme activity designed and conducted an investigation into the effect of temperature on the enzyme diastase. Diastase is a plant amylase, a starch-digesting enzyme that breaks down starch molecules into simple sugars (glucose monomers). Iodine is an effective indicator used to identify the presence or absence of starch. Iodine turns a deep blue/black colour in the presence of starch, but remains yellow/brown when starch is not present.

Before proceeding, the students tested the iodine reaction with distilled water, starch and diastase and presented the results in the following table.

Iodine reaction with:		
distilled water	**starch**	**diastase solution**
yellow	blue/black	yellow

The students set up several pairs of test-tubes containing diastase with tap water and diastase with starch solution. Each pair of test-tubes was immersed into a water bath set to a specific temperature. The colours of the test-tube solutions were recorded at time zero and again after 60 minutes (see the following table).

Temp (°C)	**Test-tube**	**Time (mins)**	
		0	**60**
0	1A: water + starch	blue/black	blue/black
	1B: diastase + starch	blue/black	blue/black
20	2A: water + starch	blue/black	blue/black
	2B: diastase + starch	blue/black	brown
30	3A: water + starch	blue/black	blue/black
	3B: diastase + starch	blue/black	yellow
40	4A: water + starch	blue/black	blue/black
	4B: diastase + starch	blue/black	brown/grey
60	5A: water + starch	blue/black	blue/black
	5B: diastase + starch	blue/black	blue/black

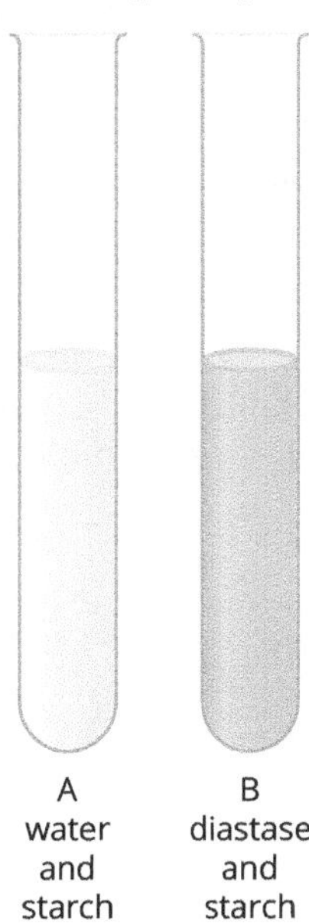

a Suggest a hypothesis for this investigation.

b Explain the reasoning behind the initial test of iodine with distilled water, starch and diastase.

c Identify the independent variable in this experiment.

d Describe three ways in which this experiment was controlled.

ISBN 978 1 4886 1931 1

e Explain the results in the following test-tubes:

i 1B:

ii 3B:

iii 5B:

f Does the experimental data support or not support your hypothesis? Explain.

g Evaluate the scientific method used in this investigation. Look carefully at the summary of the procedure. Make suggestions to improve the method for a subsequent investigation.

ISBN 978 1 4886 1931 1

MODULE 2

Organisation of living things

Outcomes

By the end of this module you will be able to:

- select and process appropriate qualitative and quantitative data and information using a range of appropriate media BIO11-4
- solve scientific problems using primary and secondary data, critical thinking skills and scientific processes BIO11-6
- communicate scientific understanding using suitable language and terminology for a specific audience or purpose BIO11-7
- explain the structure and function of multicellular organisms and describe how the coordinated activities of cells, tissues and organs contribute to macroscopic processes in organisms BIO11-9

Content

ORGANISATION OF CELLS

INQUIRY QUESTION **How are cells arranged in a multicellular organism?**

By the end of this module you will be able to:

- compare the differences between unicellular, colonial and multicellular organisms by:
  - investigating structures at the level of the cell and organelle
  - relating structure of cells and cell specialisation to function
- investigate the structure and function of tissues, organs and systems and relate those functions to cell differentiation and specialisation (ACSBL055) ICT
- justify the hierarchical structural organisation of organelles, cells, tissues, organs, systems and organisms (ACSBL054) CCT

NUTRIENT AND GAS REQUIREMENTS

INQUIRY QUESTION **What is the difference in nutrient and gas requirements between autotrophs and heterotrophs?**

By the end of this module you will be able to:

- investigate the structure of autotrophs through the examination of a variety of materials, for example: (ACSBL035) ICT
  - dissected plant materials (ACSBL032)
  - microscopic structures
  - using a range of imaging technologies to determine plant structure ICT
- investigate the function of structures in a plant, including but not limited to:
  - tracing the development and movement of the products of photosynthesis (ACSBL059, ACSBL060) ICT
- investigate the gas exchange structures in animals and plants (ACSBL032, ACSBL056) through the collection of primary and secondary data and information, for example:
  - microscopic structures: alveoli in mammals and leaf structure in plants ICT L
  - macroscopic structures: respiratory systems in a range of animals ICT L

- interpret a range of secondary-sourced information to evaluate processes, claims and conclusions that have led scientists to develop hypotheses, theories and models about the structure and function of plants, including but not limited to: (ACSBL034) CCT ICT L
  - photosynthesis
  - transpiration-cohesion-tension theory
- trace the digestion of foods in a mammalian digestive system, including: ICT L
  - physical digestion
  - chemical digestion
  - absorption of nutrients, minerals and water
  - elimination of solid waste
- compare the nutrient and gas requirements of autotrophs and heterotrophs L ICT

TRANSPORT

INQUIRY QUESTION **How does the composition of the transport medium change as it moves around an organism?**

By the end of this module you will be able to:

- investigate transport systems in animals and plants by comparing structures and components using physical and digital models, including but not limited to: (ACSBL032, ACSBL058, ACSBL059, ACSBL060) ICT L
  - macroscopic structures in plants and animals
  - microscopic samples of blood, the cardiovascular system and plant vascular systems ICT
- investigate the exchange of gases between the internal and external environments of plants and animals ICT L
- compare the structures and function of transport systems in animals and plants, including but not limited to: (ACSBL033) L
  - vascular systems in plants and animals
  - open and closed transport systems in animals
- compare the changes in the composition of the transport medium as it moves around an organism

Key knowledge

Organisation of cells

Many organisms consist of a single cell. These are called **unicellular** organisms, and include *Amoeba*, *Paramecium* and bacteria. Some organisms consist of a mass of cells that are capable of living independently but have evolved to live together in a group or **colony**.

Multicellular organisms are complex in structure as they are composed of many different kinds of cells. These different kinds of cells are organised into tissues, organs and systems. In this way, all life-sustaining functions of the organism can be met.

- A group of cells of the same type that work together to achieve a particular function is called a **tissue**. For example, a group of muscle cells is called muscle tissue.
- A group of tissues that work together to perform an overall function is called an **organ**. For example, the stomach is composed of muscle tissue and vascular tissue.
- A group of organs that work together to perform an overall function form a **system**. For example, the digestive system is composed of the mouth, stomach, liver and intestine.

Multicellular organisms may be composed of many systems, for example the digestive, respiratory, excretory and nervous systems.

CELL DIFFERENTIATION

The earliest cells in the growth and development of multicellular organisms are unspecialised. Unspecialised or undifferentiated cells are called **stem cells**—they have the potential to develop into many different cell types. As cells mature they develop the features that allow them to take on specialised roles. This is called **cell differentiation**. The structure and other features of cells are related to cell function. For example, cells lining the small intestine have microscopic finger-like projections (microvilli) on their surface that increase the surface-area-to-volume ratio of the cells, making them efficient at absorbing nutrients.

CELL SPECIALISATION

Unicellular organisms are relatively simple in structure. As the entire organism is composed of a single cell, the one cell must undertake all the functions required to sustain life. For example, the uptake of nutrients and removal of wastes for unicellular life forms may occur by diffusion directly with the external environment. This is efficient, as unicellular organisms have a high surface-area-to-volume ratio. However, this is not the case for large multicellular organisms—most of their cells are not in direct contact with the environment, so simple exchange of materials cannot occur in the same way. Specialised cells undertake many life-sustaining functions in multicellular organisms and have features that make them well suited to a particular role. Specialised cells are involved in the exchange of oxygen and carbon dioxide, transport of materials, communication and movement, to mention a few. Table 2.1 illustrates some specific examples.

TABLE 2.1 Some different types of specialised cells in plants and animals, their appearance and function

Function carried out by specialised cells	Plant cells	Animal cells
Exchange	root hair	gut epithelium cells
Transport	companion cell, sieve cell	red blood cells
Strength/support	fibres (xylem, phloem)	bone cells, cartilage cells
Protection/defence	epidermal cells, cuticle	ciliated epithelium cells, white blood cells
Photosynthesis	mesophyll cells	
Movement		muscle cells
Communication		nerve cell

ISBN 978 1 4886 1931 1

Autotroph and heterotroph nutrient and gas requirements

Living organisms have similar basic needs, but may differ in how these needs are met (Figure 2.1). **Autotrophs** produce the organic compounds they need from simple inorganic ones they obtain from the environment; for example, plants do this in the process of photosynthesis. The incorporation of carbon into organic molecules during photosynthesis is called **carbon fixation**. Chemosynthetic bacteria such as **methanogens** are another example of autotrophs. **Heterotrophs** acquire their organic substances by consuming other organisms. Organic compounds such as glucose can then be used for cellular respiration to make energy available for cells in the form of **adenosine triphosphate (ATP)**. All organisms require ATP to be able to carry out life-sustaining processes. Table 2.2 and Figure 2.2 summarise some of the key nutrient requirements of autotrophs and heterotrophs.

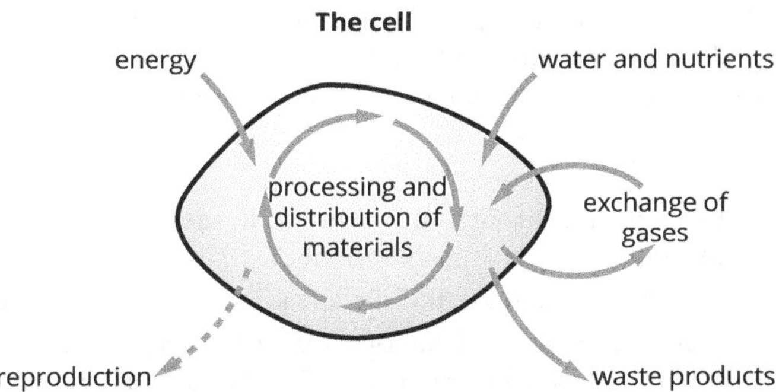

FIGURE 2.1 Common requirements of cells

TABLE 2.2 Nutritional requirements of autotrophs and heterotrophs

Nutrient	Autotrophs (plants)	Heterotrophs
Carbohydrates lipids amino acids (for proteins)	make own glucose; able to convert to other organic compounds	obtain carbohydrates from diet; nine essential amino acids must be obtained from diet, remainder can be synthesised by cells
Vitamins	able to synthesise	essential vitamins needed in diet; some can be synthesised
Minerals	obtain macronutrients such as N, P, K, Ca, Mg and S from soil	obtain minerals from diet, including Ca, P, Mg, Fe, Na, K, I and others in trace amounts
Gases	obtain O_2 and CO_2 from air and soil	obtain O_2 from air or water, depending on environment in which they live
Water	obtain from soil	terrestrial organisms obtain water from diet; aquatic organisms obtain directly from environment

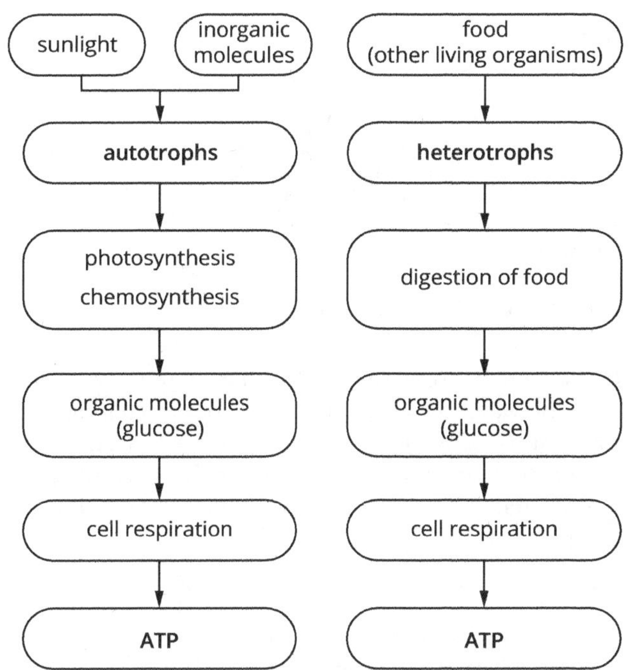

FIGURE 2.2 Comparing autotrophs and heterotrophs

GAS EXCHANGE IN PLANTS

Gas exchange structures in plants are relatively simple compared to the complex structures that are present in multicellular animals. Openings in leaves, **stomata**, allow **carbon dioxide** (for photosynthesis) and **oxygen** (for cellular respiration) to enter the leaf, where gas exchange occurs directly with cells by diffusion. The opening and closing of stomata is controlled by a pair of guard cells (Figures 2.3 and 2.4). Cells are arranged in a loosely packed fashion that allows diffusion to occur more rapidly (Figure 2.3). The size of the stomatal aperture is regulated by the turgidity of the guard cells. When the guard cells are turgid—under water pressure—they swell and buckle and the stoma becomes larger. A loss of turgor leads to guard cells straightening and the stomatal pore closes.

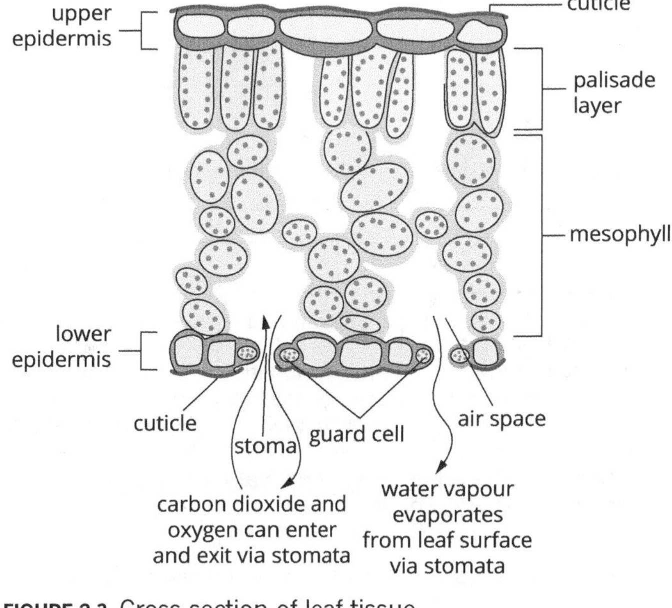

FIGURE 2.3 Cross section of leaf tissue

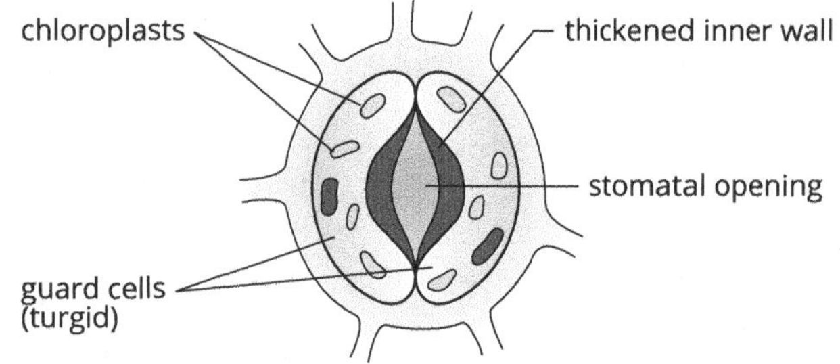

FIGURE 2.4 Stomata and guard cells

Gas exchange in plants needs to be balanced with conserving water. While stomata are open, water vapour can be lost to the environment. This poses problems for plants that live in dry environments. To overcome this, arid-adapted plants only open their stomata during cooler parts of the day.

GAS EXCHANGE IN ANIMALS

While the process of gas exchange in animals is always by diffusion between the external and internal environment, the structures involved to facilitate this vary between different kinds of animals (Figure 2.5). Oxygen and carbon dioxide need to be exchanged continuously for an organism to function efficiently. Unicellular and some very small organisms exchange these gases directly with the environment by diffusion. Large, multicellular organisms feature specialised respiratory surfaces to ensure adequate gas exchange.

Efficient respiratory surfaces feature:

- a large surface area
- thin, moist and easily penetrable surfaces
- adequate ventilation
- efficient transport of carrier fluid (blood) across the respiratory surface.

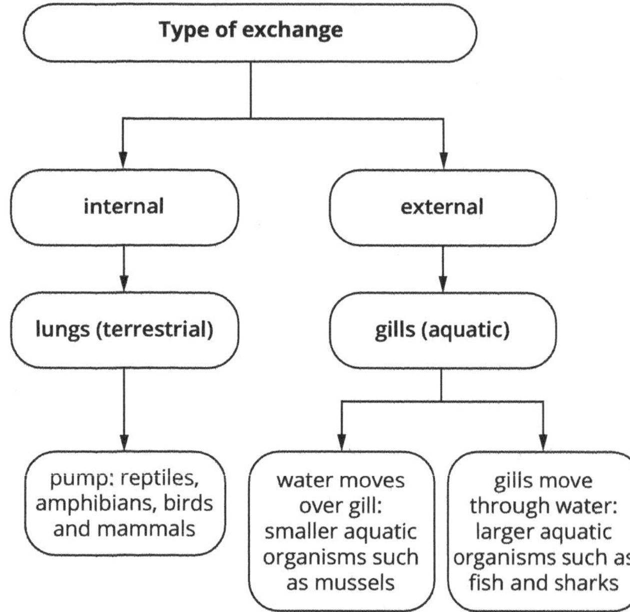

FIGURE 2.5 Gas exchange systems vary according to an organism's environment.

Gas exchange systems of complex organisms are related to the environment in which they live. Obtaining oxygen from the air or water has both advantages and disadvantages (Table 2.3).

TABLE 2.3 Comparing water and air ventilation

	Water ventilation	Air ventilation
Main respiratory organ	gills	lungs
Advantages	• no drying out of respiratory surface • carbon dioxide easily diffuses into surrounding water • countercurrent flow of blood increases rate of gas exchange	• less energy required • more oxygen in air
Disadvantages	• more energy required to ventilate gills • less oxygen in water	• loss of water from respiratory surface by evaporation during exhalation

Gas exchange in mammals

The mammalian respiratory system features structures and processes that result in the efficient exchange of oxygen and carbon dioxide. It is composed of:

- **lungs**
- respiratory passages—nasal passages, trachea, bronchi, and bronchioles that end in sacs called **alveoli** (the site of gas exchange).

The alveoli increase the surface area for gas exchange.

Contraction and relaxation of the diaphragm muscle results in changes in pressure within the lung cavity, forcing air in and out.

The epiglottis prevents food from travelling down the trachea and blocking the passage of air.

Gas exchange relies on diffusion, therefore **concentration gradients** need to be maintained. For oxygen to continually diffuse into **capillaries** around alveoli, it needs to be transported away from the site. Respiratory pigments such as **haemoglobin** carry oxygen in the blood from the lungs and release oxygen into cells. Rhythmic pumping of the heart ensures the continuous flow of blood along capillaries surrounding the alveoli.

Haemoglobin and oxygen bind in a reversible reaction:

$$\text{Hb} + 4\text{O}_2 \rightleftharpoons \text{Hb(O}_2\text{)}_4$$
$$\text{haemoglobin} + \text{oxygen} \rightleftharpoons \text{oxyhaemoglobin}$$

The reaction proceeds to the right in areas of high oxygen concentration, such as when oxygen diffuses into the capillaries in the lungs. The reaction proceeds to the left in areas of low oxygen concentration, such as in capillaries near body cells. The released oxygen can now diffuse into nearby cells. Diffusion of each gas occurs down its concentration gradient, whether it is in the alveoli of the lungs or in capillaries near body cells (Figure 2.6).

ISBN 978 1 4886 1931 1

inhaled air
alveolus
carbon dioxide molecule
oxygen molecule
$Hb + 4O_2 \rightarrow Hb(O_2)_4$
blood vessel in lungs
blood vessel near cells
$Hb + 4O_2 \leftarrow Hb(O_2)_4$
body cell

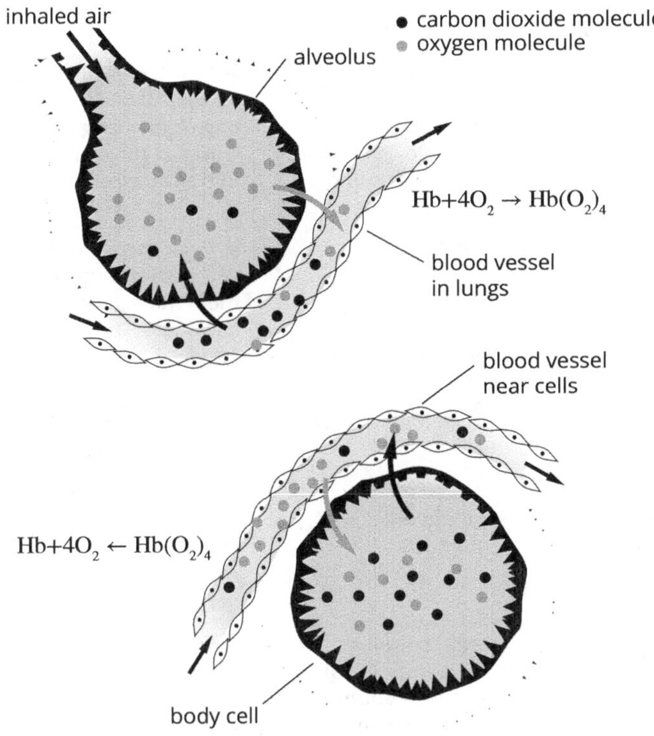

FIGURE 2.6 Diffusion of gases in the lungs and body cells

OBTAINING NUTRIENTS: HETEROTROPH DIGESTIVE SYSTEMS

The foods consumed by heterotrophs are useful on a cellular level. However, most organic nutrients ingested in the diet are in a form that is too large to enter cells. This organic matter needs to be broken down into simpler forms so that it can ultimately pass across cell membranes and into cells. This process of breaking large food molecules into smaller ones is called **digestion**.

Different kinds of heterotrophs have different digestive structures depending on their diets (Figure 2.7).

Larger, complex animals need to ensure digestion occurs efficiently enough to meet generally higher energy demands. These organisms have the following characteristics:

- adaptations for acquiring food
- reliance on the physical breakdown of food
- appropriate structures for handling and mastication of food
- a one-way gut compartmentalised into different sections, each providing an optimal environment for different digestive enzymes
- a large surface area for absorption of nutrients from the gut into the circulatory system
- movement of food along the digestive system and removal of undigested material.

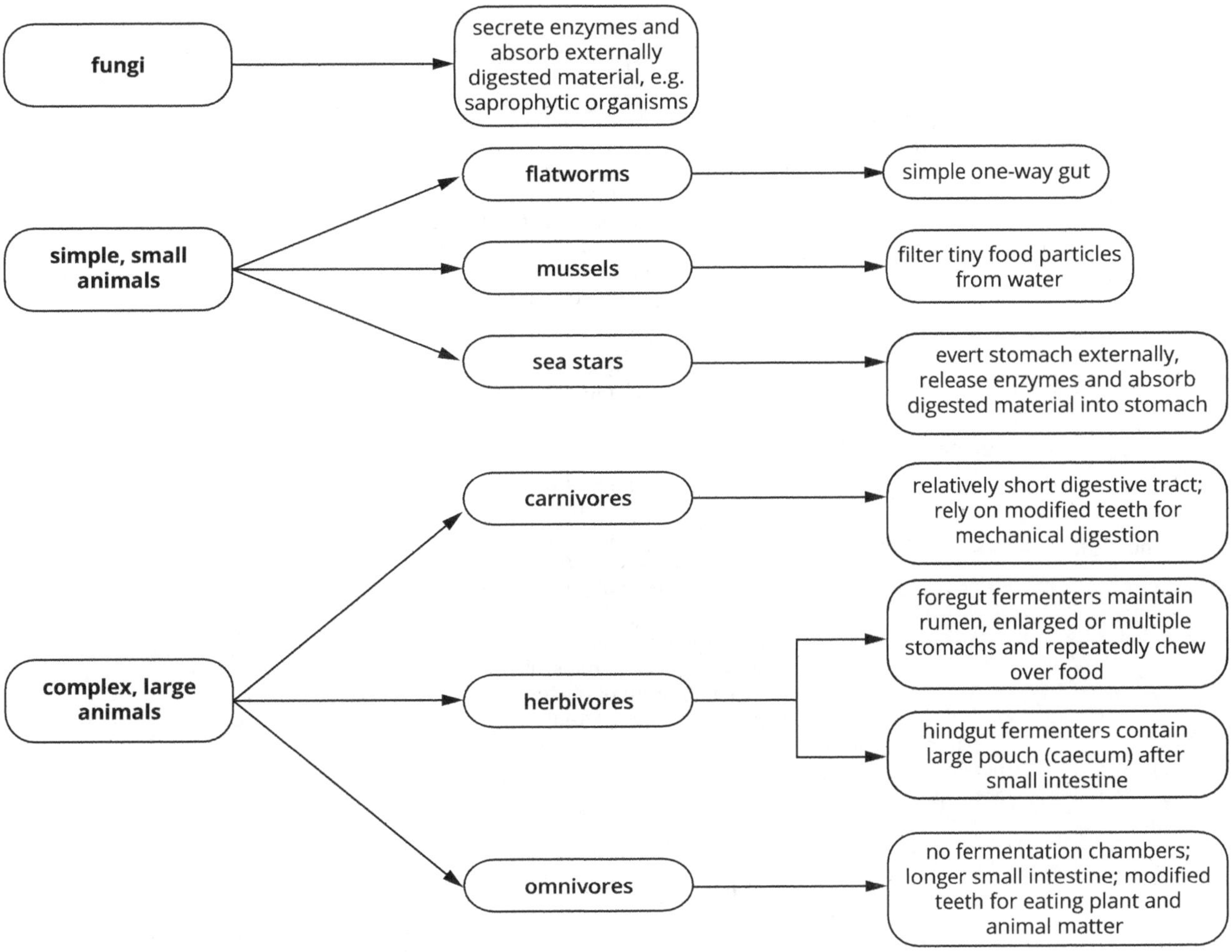

fungi
secrete enzymes and absorb externally digested material, e.g. saprophytic organisms
simple, small animals
flatworms
simple one-way gut
mussels
filter tiny food particles from water
sea stars
evert stomach externally, release enzymes and absorb digested material into stomach
complex, large animals
carnivores
relatively short digestive tract; rely on modified teeth for mechanical digestion
herbivores
foregut fermenters maintain rumen, enlarged or multiple stomachs and repeatedly chew over food
hindgut fermenters contain large pouch (caecum) after small intestine
omnivores
no fermentation chambers; longer small intestine; modified teeth for eating plant and animal matter

FIGURE 2.7 Outline of different digestive systems among heterotrophs

Feeding behaviours vary between **herbivores**, **carnivores** and **omnivores**. The digestive systems of these three groups are suited to their diet and behaviour. In particular, herbivores such as rabbits, cows and koalas rely on the presence of certain bacteria in their digestive tracts to provide cellulase, the enzyme necessary to break down the cell walls (cellulose) of plant matter. Omnivores and carnivores are unable to digest plant matter to the same extent as herbivores.

Mammalian digestive systems

Simple animals rely primarily on chemical processes involving enzymes for digestion. Larger animals (humans, for example) also utilise mechanical processes during digestion (Table 2.4). Mammalian digestive systems primarily consist of:

- mouth
- oesophagus
- stomach
- small intestine
- large intestine
- rectum
- anus
- associated organs such as the salivary glands, liver, gall bladder and pancreas that release enzymes and other chemicals into the digestive tract (Figure 2.8).

TABLE 2.4 Comparison of physical and chemical digestion

	Physical digestion	Chemical digestion
Type of process	• mechanical breakdown of large food pieces into smaller pieces	• chemical change of complex molecules into simpler ones
Purpose	• smaller pieces have a relatively large surface area that allows enzymes to work more effectively	• simpler molecules can pass through cell membranes and into cells
Requires	• teeth and muscles	• enzymes and appropriate pH conditions
Where it occurs	• mouth and stomach	• mouth, stomach and small intestine • rumen and caecum in herbivores
Examples	• teeth chewing food, muscles of stomach churning food into semi-liquid paste called chyme	• lipases act on lipids • proteases act on proteins • amylases act on carbohydrates • gastric lipase begins converting fats into fatty acids and glycerol

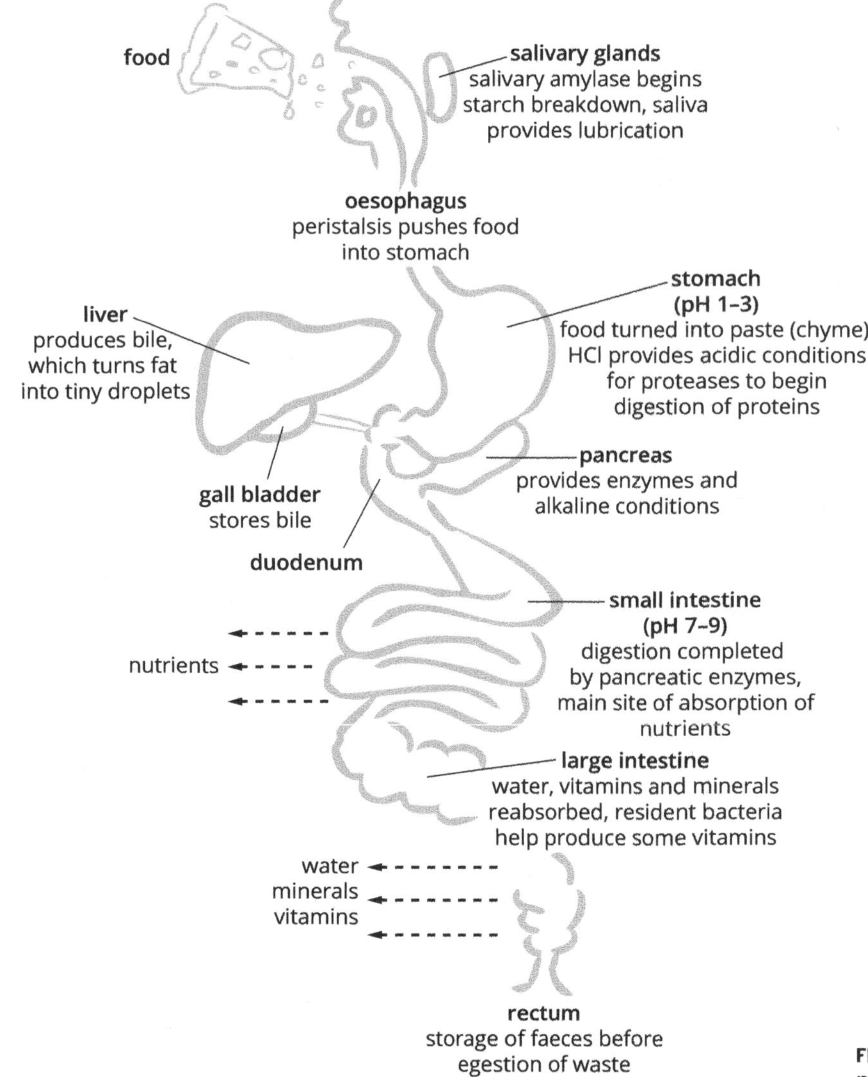

FIGURE 2.8 Schematic diagram of digestive processes in mammals

ISBN 978 1 4886 1931 1

Transport

TRANSPORT IN PLANTS

Plants feature a series of tubes called vascular tissue for the transport of substances. There are two key types of vascular tissue in plants: **xylem** and **phloem**.

- Xylem—composed of non-living tissue; remnants of cells reinforced with lignin.
- Phloem—living cells; end plates of cells sieve-like; associated with companion cells that control activities of nucleus-free sieve cells.

Plasmodesmata are fine channels that link plant vascular cells, allowing lateral movement of nutrients from cell to cell.

Transpiration is the loss of water from plants through evaporation, mainly via stomata. The cohesion of water molecules means that as water vapour is evaporated at the leaf surface, the entire water column is drawn up through the xylem.

Translocation refers to the movement of organic solutes from the leaves, where they are produced during photosynthesis, to other parts of the plant.

Table 2.5 summarises the roles of xylem and phloem in transport in plants.

TABLE 2.5 Characteristics of transport in plants

Vascular tissue	Xylem	Phloem
Substances involved	• water • inorganic nutrients	• organic nutrients, e.g. sugars
Direction of transport	• from roots up through the plant	• from leaves to rest of plant in both directions (upwards and downwards)
Processes involved	• transpiration and root pressure draw water upwards • no energy expenditure by the plant	• active process requiring energy

TRANSPORT IN ANIMALS

Unicellular organisms, due to their size (and high surface-area-to-volume ratio), have no need for specialised systems to deliver nutrients to, and remove wastes from, all parts of the organism. Multicellular organisms rely on complex transport systems to ensure the needs of cells are met. Animal transport systems can be **open circulatory systems** or **closed circulatory systems**. Table 2.6 summarises the key features of these.

The circulatory systems of complex multicellular animals, such as mammals, are characterised by:

- an efficient carrying capacity
- a large surface area for the exchange of materials
- the effective movement of fluid
- the ability to regulate movement of fluid according to needs.

TABLE 2.6 Circulatory systems in animals

Transported by	Description	Examples
Diffusion across body surface	• structure of body provides adequate surface area, e.g. flat, thin	• flatworms, sea kelps
Open circulatory system	• body fluid around cells is circulated using muscular contraction • low pressure and slow cycle	• insects, spiders, snails
Closed circulatory system	• transporting fluid (blood) enclosed in system of vessels • high pressure and fast cycle • could be – single circuit heart → gills → body – double circuit heart → lungs → heart → body	• fish and most molluscs • all vertebrates (except fish) • squid and octopus

Closed circulatory systems involve movement of blood that is separated from the cells of the body and the interstitial fluid.

Blood contains both intracellular and extracellular fluid and is surrounded by interstitial fluid where diffusion takes place between capillaries and body cells (Figure 2.9).

- **Intracellular fluid:** fluid inside cells (cytosol).
- **Extracellular fluid:** all fluid outside cells.
- **Interstitial fluid:** extracellular fluid surrounding cells and separate from blood.

Mammalian transport systems

Mammalian transport systems include the lymphatic drainage system and the blood circulatory system (also known as the cardiovascular system). The lymphatic drainage system:

- contains fluid called **lymph**
- helps maintain osmotic and fluid balance in tissues
- collects leaked proteins from blood capillaries
- returns lymph to veins near heart.

The blood circulatory system:

- is a closed system
- has a double circuit involving pulmonary and systemic circulation
- consists of blood vessels—**arteries** (carry blood away from heart), **veins** (carry blood to heart) and **capillaries** (smallest blood vessels; site of exchange)
- has a muscular pump called a **heart**.

Pulmonary circulation is the transport of deoxygenated blood from the right side of the heart to the lungs and the return of oxygenated blood back to the left side of the heart (Figure 2.10).

Systemic circulation is the transport of oxygenated blood from the left side of the heart to the rest of the body and the return of deoxygenated blood back to the right side of the heart (Figure 2.10).

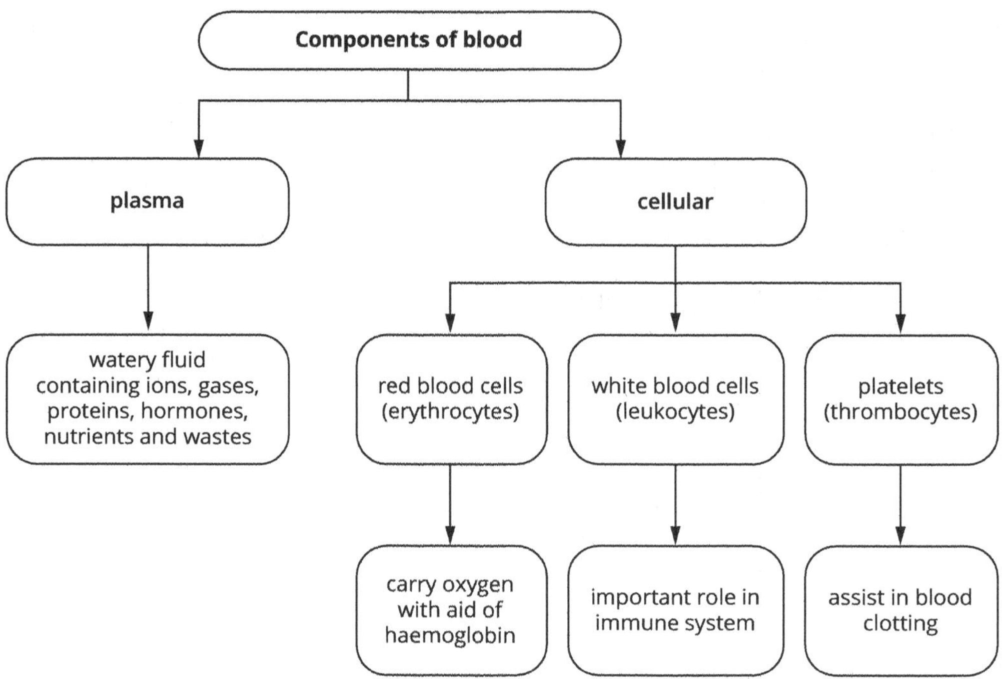

FIGURE 2.9 Components of blood

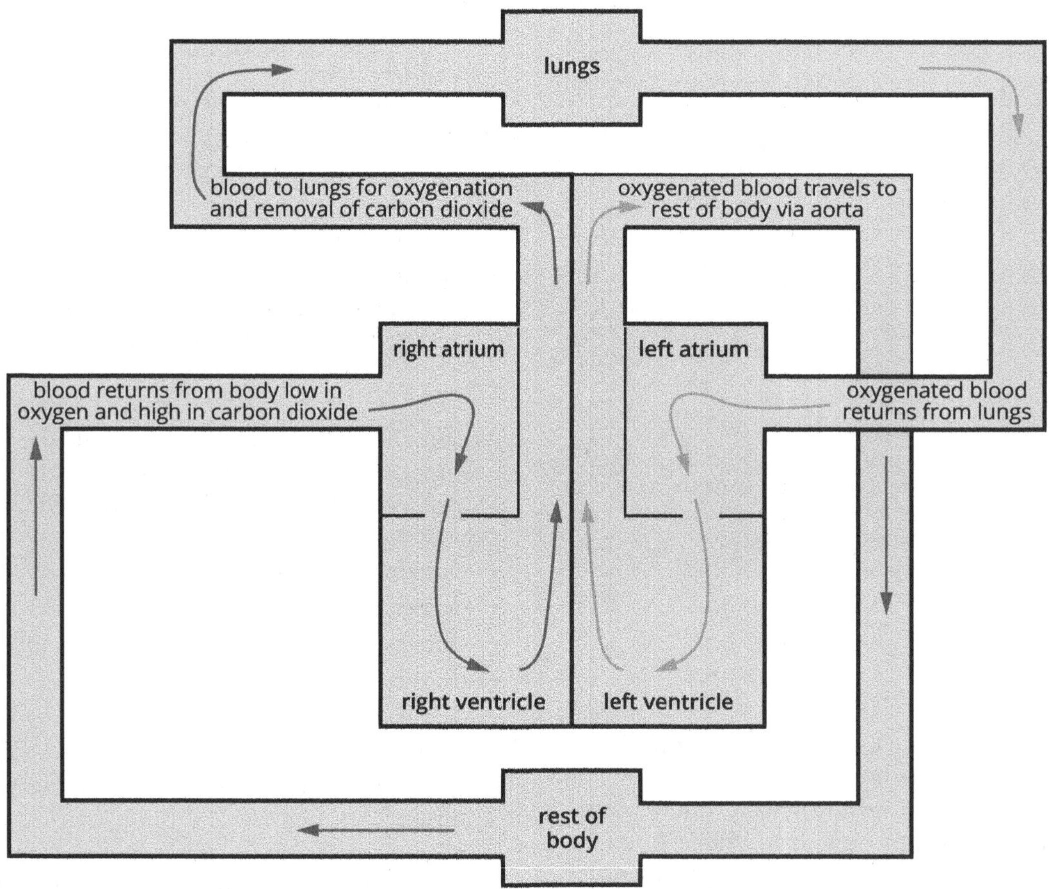

FIGURE 2.10 Schematic diagram showing circulation in mammals

ISBN 978 1 4886 1931 1

WORKSHEET 2.1

Knowledge review—revisiting structure and function of organisms

1 Following are some key terms that you have studied before in biology and which you will encounter in this module. This worksheet is designed to familiarise yourself with them. Read the definitions provided in the boxes at the right. Choose the correct term from the list below to match each definition.

multicellular	herbivore	respiratory system	tissue	digestive system	roots	skin
omnivore	organ	circulatory system	carnivore	unicellular	specialised cell	leaf

	an organism whose diet is composed primarily of the flesh of other animals
	a cell that has features that make it well suited to a particular role
	a plant organ with a structure that makes it specialised for harnessing sunlight in the process of photosynthesis
	a group of different kinds of tissues that work together to perform a specialised overall function
	describes an organism that is composed of many cells
	a body system in animals that is responsible for gas exchange
	an organism whose diet is composed primarily of plant material
	describes an organism that is composed of a single cell
	a system in many animals that is responsible for the breakdown of foods into smaller particles that can then be absorbed into the bloodstream
	the largest organ in the human body, provides a barrier between the internal and external environment, also involved in other roles, e.g. temperature regulation

2 There are four terms remaining. For each of these identify whether it is a plant or animal structure and write a sentence describing its role.

Term 1:

Term 2:

Term 3:

Term 4:

WORKSHEET 2.2

Organised organisms—levels of organisation in multicellular organisms

Use your knowledge and understanding of the levels of organisation in multicellular organisms to complete the following table. In each case, write the definition and list three other examples.

Item	Definition	Specific example	Other examples
Specialised cell		muscle cell	• • •
Tissue		muscle tissue	• • •
Organ		heart	• • •
System		circulatory system	• • •
Organism		human body	• • •

RATING MY LEARNING	My understanding improved	Not confident ⟵⟶ Very confident ○ ○ ○ ○ ○	I answered questions without help	Not confident ⟵⟶ Very confident ○ ○ ○ ○ ○	I corrected my errors without help	Not confident ⟵⟶ Very confident ○ ○ ○ ○ ○

ISBN 978 1 4886 1931 1

WORKSHEET 2.3

Cells on a mission—cell specialisation

1 Consider the examples of specialised cells shown in the following table. For each cell type shown, suggest its function and outline the features that make it well suited to this role. Also, identify the cell types represented by b–e.

	Cell type	Diagram	Function	Feature
a	nerve cell	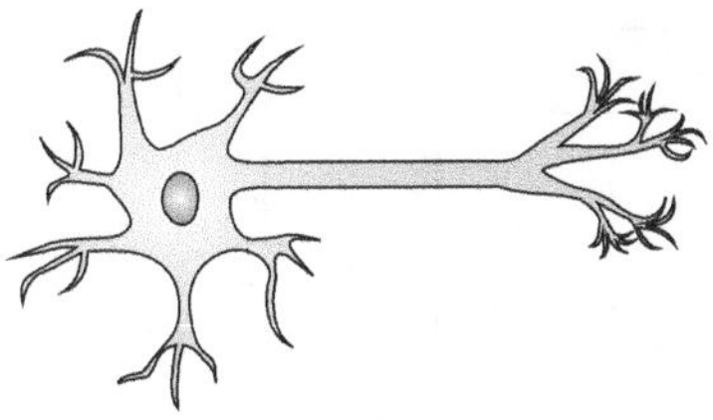		
b		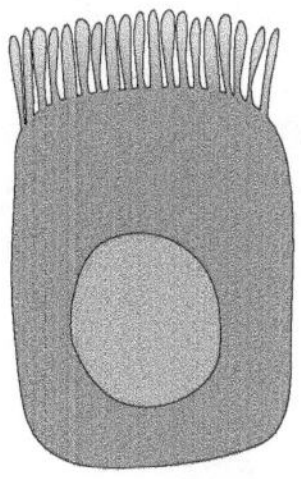		
c		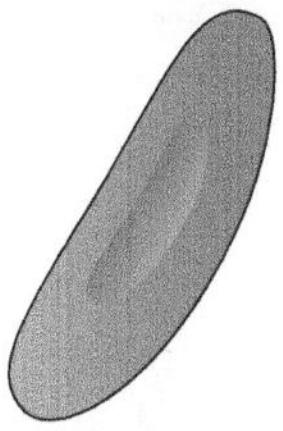		
d		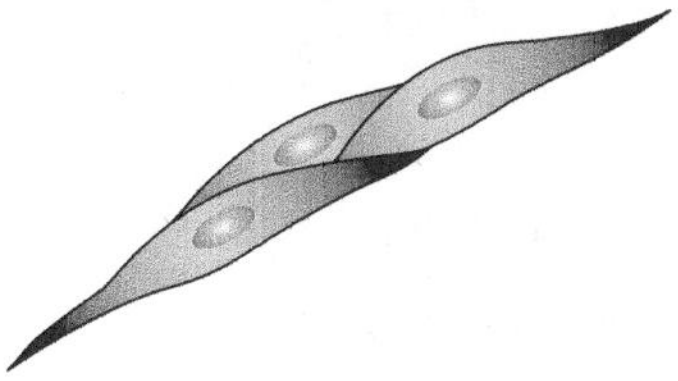		
e		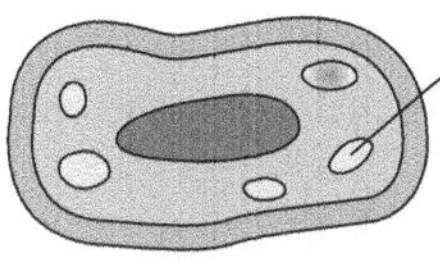 chloroplast		
f	root hair cell	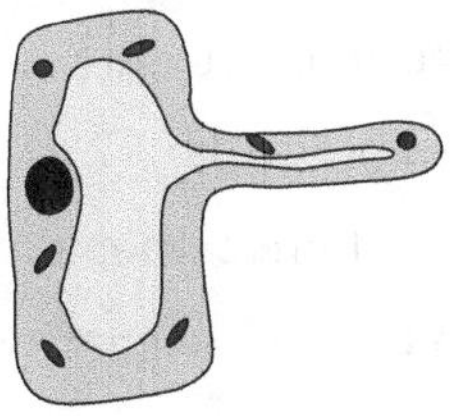		

2 Summarise the relationship between the structure and features of cells and their function.

RATING MY LEARNING	My understanding improved	Not confident ◄——► Very confident ○ ○ ○ ○ ○	I answered questions without help	Not confident ◄——► Very confident ○ ○ ○ ○ ○	I corrected my errors without help	Not confident ◄——► Very confident ○ ○ ○ ○ ○

WORKSHEET 2.4

Crossword—autotrophs and heterotrophs

Complete the crossword to check your knowledge and understanding of the requirements of autotrophs and heterotrophs.

Across

7 Organism that produces its own energy-rich compounds from inorganic ones

8 Photosynthetic pigment contained inside the chloroplasts of green plants

11 Process in which some autotrophic bacteria produce organic compounds from inorganic ones

13 Organelle that is the site of photosynthesis

14 Process in which autotrophic organisms incorporate carbon atoms into organic molecules

Down

1 A special kind of heterotroph that obtains its nutrients by secreting breakdown enzymes onto the dead and decaying matter on which it grows, before absorbing the digested material

2 A group of chemosynthetic bacteria that inhabit extreme environments

3 A chemosynthetic bacterium that produces methane (organic) from carbon dioxide and hydrogen

4 Process in which the cells of organisms break down glucose to make energy available

5 Process in which green plants harness sunlight to produce organic compounds from inorganic ones

6 Describes relatively simple molecules that do not contain carbon and hydrogen

9 Describes an energy-rich carbon and hydrogen containing compound

10 Organism that does not make its own organic compounds, feeding instead on other organisms to acquire these

12 Energy-rich molecule that is both produced in photosynthesis and a raw material for cellular respiration

RATING MY LEARNING	My understanding improved	Not confident ◄——► Very confident ○ ○ ○ ○ ○	I answered questions without help	Not confident ◄——► Very confident ○ ○ ○ ○ ○	I corrected my errors without help	Not confident ◄——► Very confident ○ ○ ○ ○ ○

 ISBN 978 1 4886 1931 1

WORKSHEET 2.5

Open all hours—stomata and gas exchange in plants

In plants, gas exchange between the tissues and the external environment occurs through pores in the epidermis called stomata. Each stoma is bounded by two guard cells.

1 Label the structures shown in the cross-section of a typical leaf.

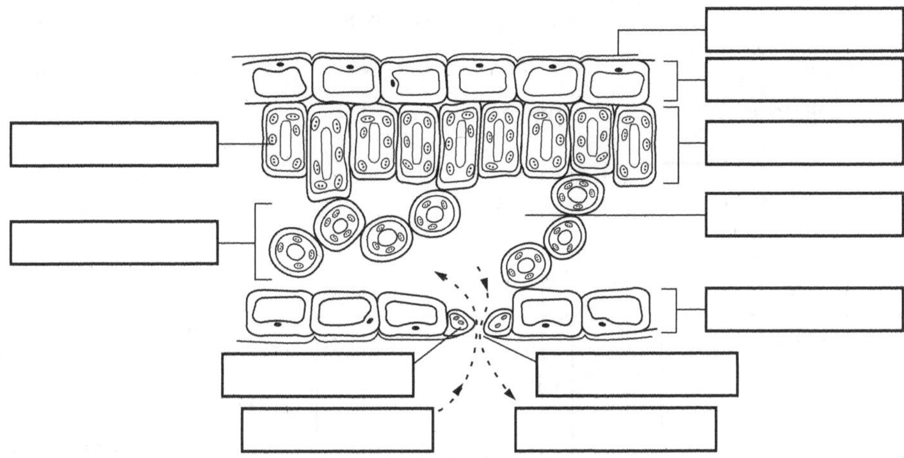

a What unusual feature do guard cells possess that is different from other epidermal cells?

b The stomata are the site of gas exchange in plants. In many species of plants the distribution of stomata between the upper and lower epidermis is uneven. Suggest an explanation for this.

2 Look carefully at the differences between the shapes of the guard cells in the following diagrams.

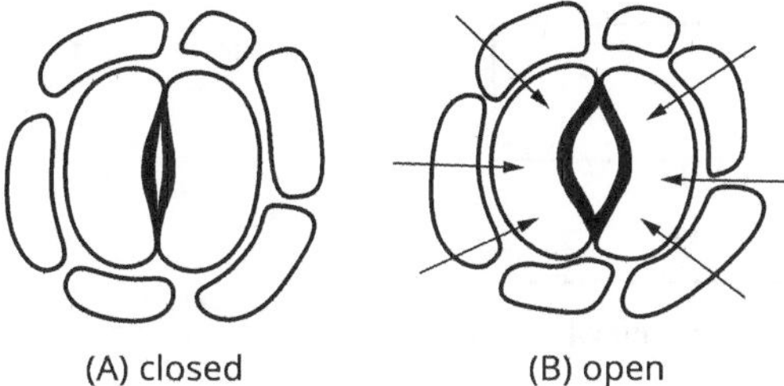

(A) closed (B) open

a Label the guard cells in A.

b The arrows in B represent an input into the cell. Suggest what the input is, and outline the effect it has had on the guard cells.

c Describe three features of guard cells that explain their control over the stomatal aperture.

Feature 1:

Feature 2:

Feature 3:

RATING MY LEARNING	My understanding improved	Not confident ◄——► Very confident ○ ○ ○ ○ ○	I answered questions without help	Not confident ◄——► Very confident ○ ○ ○ ○ ○	I corrected my errors without help	Not confident ◄——► Very confident ○ ○ ○ ○ ○

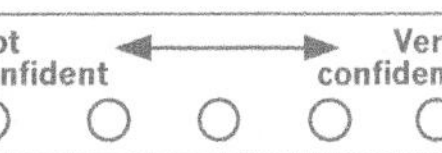

WORKSHEET 2.6

Digesting dinner—heterotroph digestion

Use the key words listed below to complete the concept map summarising nutrition and digestion in heterotrophs. Write along the link lines between terms and phrases to show the relationships between ideas in your concept map.

amino acids	egestion	chemical digestion	caecum	carbohydrates	duodenum	rectum
cellulase	fermentation	bile	ileum	carnivore	proteins	villi
digestive enzymes	chyme	lipids	vitamins	herbivore	rumen	digestion
physical breakdown	glycogen	cellulose	omnivore	colon		

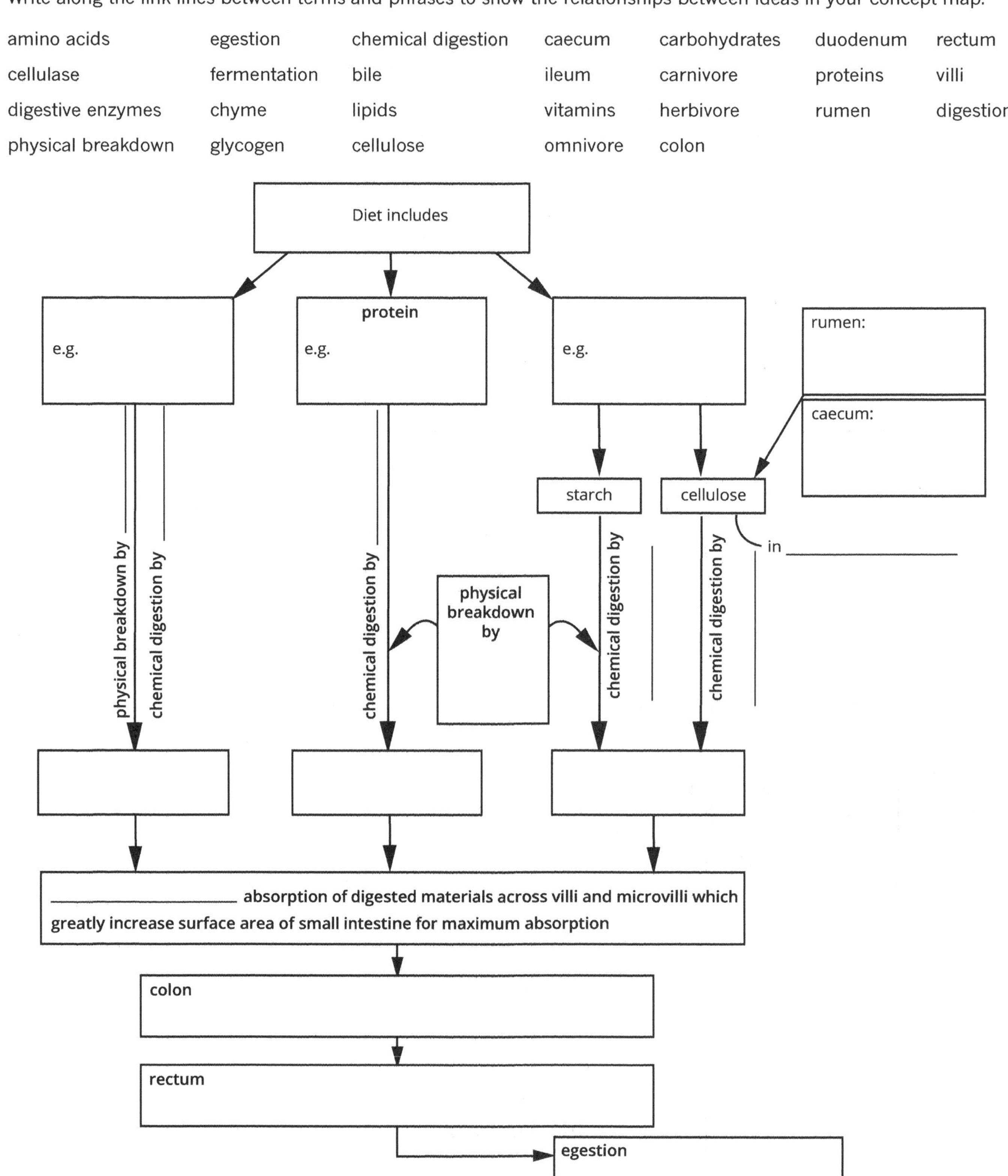

RATING MY LEARNING				
My understanding improved	Not confident ◄—► Very confident	○ ○ ○ ○ ○		
I answered questions without help	Not confident ◄—► Very confident	○ ○ ○ ○ ○		
I corrected my errors without help	Not confident ◄—► Very confident	○ ○ ○ ○ ○		

ISBN 978 1 4886 1931 1

WORKSHEET 2.7

Gas exchange—human respiratory system

1 Select terms from the list below to complete the summary statements, outlining the key points about gas exchange in humans.

alveolar sacs	bronchiole	cellular respiration	diaphragm	trachea
ventilation	diffusion	alveoli	bronchus	oxyhaemoglobin

The exchange of gases across respiratory surfaces in aerobic organisms makes oxygen available for ______________ ______________ and ensures that carbon dioxide is removed.

Gas exchange at respiratory surfaces occurs by the process of ______________.

In animals, the movement of air or water across the respiratory surfaces is called ______________.

Terrestrial animals such as mammals ventilate lungs to achieve gas exchange. A series of respiratory passages lead from the nasal passages to the lungs, where gas exchange occurs at the ______________.

The oxygen-carrying molecule in mammals is haemoglobin. When oxygen combines with this molecule in the red blood cells it forms a complex called ______________. Carbon dioxide is carried in the blood in dissolved form.

A muscular wall at the base of the thorax called the ______________ rhythmically expands and contracts to ventilate the lungs.

2 Use the remaining terms to label the following diagram that illustrates the pathway that air takes as it enters and leaves the body, allowing exchange of gases at the respiratory surface.

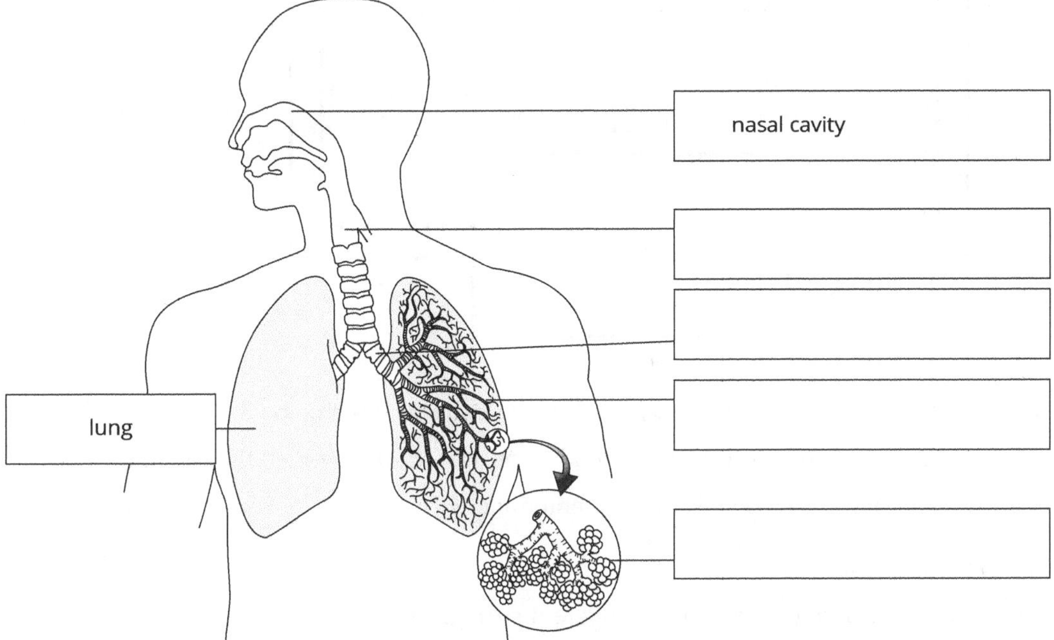

3 Describe the properties of gas exchange surfaces and explain how each property assists in the efficient diffusion of gases.

____

____

____

____

RATING MY LEARNING	My understanding improved	Not confident ◄──► Very confident ○ ○ ○ ○ ○	I answered questions without help	Not confident ◄──► Very confident ○ ○ ○ ○ ○	I corrected my errors without help	Not confident ◄──► Very confident ○ ○ ○ ○ ○

ISBN 978 1 4886 1931 1

WORKSHEET 2.8

Lifeline—the mammalian circulatory system

The following diagram illustrates the circuit that blood takes as it moves around a mammalian body. Follow the instructions below to complete the diagram to produce an information chart about the circulatory system of mammals.

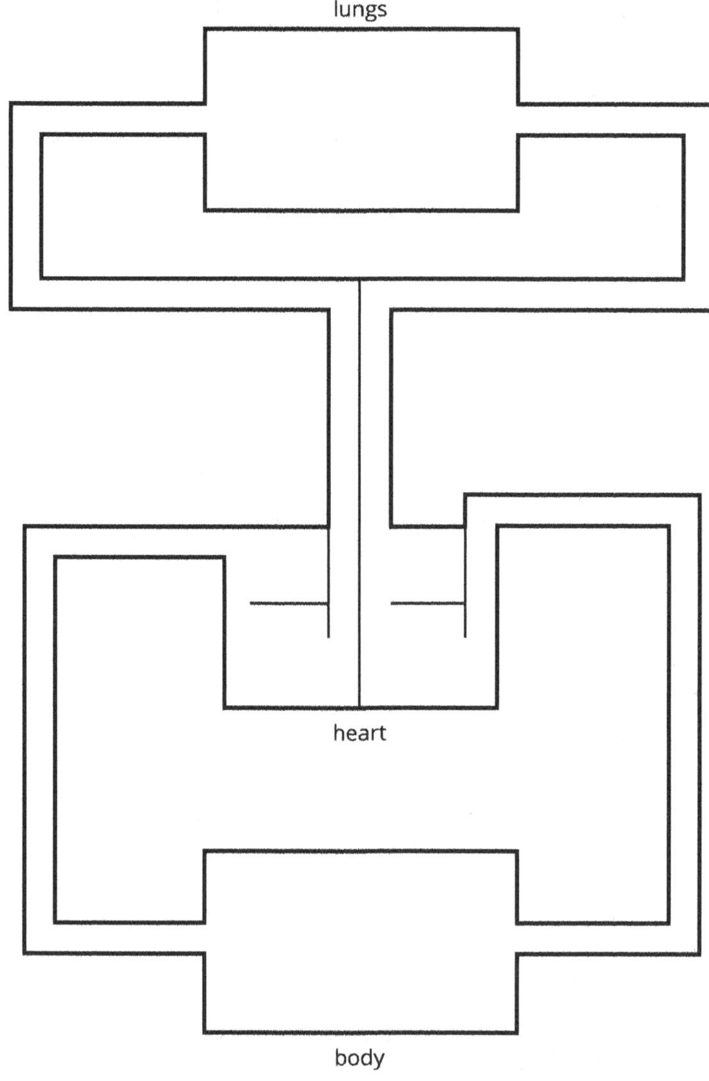

1 Colour code the diagram to show the movement of oxygenated and deoxygenated blood.

2 Insert arrows to indicate the direction and pathway that blood takes as it moves through the circulatory system.

3 Label the chambers of the heart and the blood vessels that enter and leave it.

4 Indicate the location of valves, for example:

5 Add notes that outline the features and functions of the different structures.

RATING MY LEARNING	My understanding improved	Not confident ◄——► Very confident ○ ○ ○ ○ ○	I answered questions without help	Not confident ◄——► Very confident ○ ○ ○ ○ ○	I corrected my errors without help	Not confident ◄——► Very confident ○ ○ ○ ○ ○

ISBN 978 1 4886 1931 1

WORKSHEET 2.9

Literacy review—terms for transport

1 Use an online dictionary to find the definition of the term 'vascular'. Write your own definition of the term below.

2 Write a paragraph summarising how the term vascular relates to transport systems in organisms. Your summary should make clear how the term is relevant in meeting the needs of both large, multicellular plants and animals. Include the following list of words in your summary.

vascular   transport   nutrients   wastes   organisms   vessels   large, multicellular organisms

3 Draw a mind map to summarise your knowledge and understanding of the following terms relevant to transport in large, multicellular plants and animals. Think carefully about the flow of ideas, writing brief notes along the link lines that show the relationship between terms. Your mind map should be easy for a classmate to read and understand.

vascular tissue   open transport system   phloem   closed transport system   xylem   transpiration   cells

RATING MY LEARNING	My understanding improved	Not confident ◄—► Very confident ○ ○ ○ ○ ○	I answered questions without help	Not confident ◄—► Very confident ○ ○ ○ ○ ○	I corrected my errors without help	Not confident ◄—► Very confident ○ ○ ○ ○ ○

WORKSHEET 2.10

Thinking about my learning

On completion of Module 2: Organisation of living things, you should be able to describe, explain and apply the relevant scientific ideas. You should be able to work with data, to interpret, analyse and evaluate it.

1 The following table lists the key knowledge covered in this module. Read each and reflect on how well you understand each concept. Rate your learning by shading the circle that corresponds to your level of understanding for each concept. It may be helpful to use colour as a visual representation. For example:

- green—very confident
- orange—in the middle
- red—starting to develop.

Concept focus	**Rate my learning** Starting to develop ◄——► Very confident				
Unicellular, colonial and multicellular organisms	○	○	○	○	○
Specialised cells in terms of their structure and function	○	○	○	○	○
Levels of organisation in living things—cells, tissues, organs, systems, organism	○	○	○	○	○
Requirements of autotrophs and heterotrophs	○	○	○	○	○
Structure and function of gas exchange systems in plants	○	○	○	○	○
Structure and function of gas exchange systems in animals	○	○	○	○	○
Structure and function of digestive systems in animals	○	○	○	○	○
Structure and function of transport systems in plants	○	○	○	○	○
Structure and function of transport systems in animals	○	○	○	○	○

2 Consider points you have shaded from starting to develop to middle-level understanding. List specific ideas you can identify that were challenging.

____

____

____

3 Write down two different strategies that you will apply to help further your understanding of these ideas.

____

____

____

ISBN 978 1 4886 1931 1

PRACTICAL ACTIVITY 2.1

Cell specialisation—microscopy

Suggested duration: 50 minutes

INTRODUCTION

In this activity you will use a light microscope to view a range of tissues from different kinds of organisms. While the cells of all organisms share some common features, there are also differences between the cells of plants and those of animals. You will look for the common features between cells and the differences between the cells of different kinds of tissues, as well as the similarities observed in the tissues of plant cells and the similarities observed in the tissues of animal cells. Observing these patterns is helpful in understanding the overall structure and organisation of cells in different groups of organisms.

MATERIALS

- light microscope
- selected prepared slides, for example:
  - cross-section of green plant stem (dicotyledon)
  - leaf epidermis
  - nerve cells
  - white blood cells
  - muscle tissue
  - lung tissue

PURPOSE

To observe tissues from different kinds of organisms under the light microscope to further understand cell structure and specialisation.

PROCEDURE

1. Set up the light microscope on the workbench.
2. Collect at least four prepared slides—two plant and two animal.
3. Select the green plant stem first. Place the slide on the stage of the microscope, focus using low power, before viewing under high power. For further information about viewing specimens under the light microscope, refer to the procedures on pages 5–7.
4. Prepare a detailed drawing of several of the different kinds of cells evident in the cross section. Remember to provide a title, magnification and labelling for your drawing.
5. Select a second plant specimen. Repeat step 4.
6. Select two different animal tissue specimens in turn. Draw only two or three of the cells.
7. At the conclusion of the investigation, return the microscope and the prepared slides to their storage containers.

Cell drawings

ANALYSIS OF RESULTS

1 Describe two similarities you observed in the plant tissue specimen.

2 Describe two similarities you observed in the animal tissue specimen.

3 Outline two differences you observed between the plant cells and animal cells.

4 The cells you have observed have been preserved as well as coloured using dyes of different kinds. Suggest a reason for the use of dyes in these specimens.

CONCLUSION

5 Outline any patterns you observed in this investigation.

6 Comment on the reliability of conclusions you can draw from this investigation.

RATING MY LEARNING	My understanding improved	Not confident ◀——▶ Very confident ○ ○ ○ ○ ○	I answered questions without help	Not confident ◀——▶ Very confident ○ ○ ○ ○ ○	I corrected my errors without help	Not confident ◀——▶ Very confident ○ ○ ○ ○ ○

 ISBN 978 1 4886 1931 1

PRACTICAL ACTIVITY 2.2

Food tubes—comparing digestive systems

Suggested duration: 50 minutes

INTRODUCTION

Mammals feed on a large variety of plants and animals or their products. Different mammals have evolved gut structures suited to different food diets. For example:

- vampire bats feed on blood
- pigs feed on a varied diet similar to that of humans
- sheep feed on plants that have a high proportion of cellulose, which can only be broken down by microbial action.

PURPOSE

To compare the digestive systems of a number of animals.

To make some inferences about the structure and function of the digestive system of different kinds of animals based on diet.

PROCEDURE

Examine the diagrams of the three mammalian digestive tracts shown at right.

ANALYSIS OF RESULTS

1 Comment on the relative size and shape of the stomachs of the three mammals. Explain the difference in terms of the type of food eaten and the amount of digestion that is required. Think carefully about what each mammal's diet consists of.

   a sheep

   b pig

   c vampire bat

(a) Sheep

oesophagus
stomach
caecum
large intestine
small intestine

(b) Pig

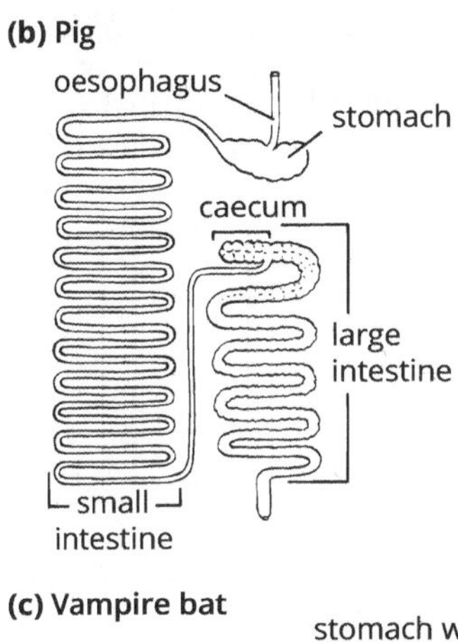

(c) Vampire bat

stomach with very flexible wall (able to expand greatly)
oesophagus
large intestine
small intestine
no caecum

Digestive tract of three mammals: (a) sheep (b) pig (c) vampire bat

PRACTICAL ACTIVITY 2.2

2 Compare the relative length of the small intestine in the three mammals. Considering the diet of the different mammals, account for any differences observed in length.

 a sheep

 b pig

 c vampire bat

3 Suggest how digestion in a sheep would be likely to be affected if its gut had the same proportions as that of the vampire bat? Would you expect it to acquire nutrients efficiently? Explain.

4 Explain why herbivores such as sheep need gut bacteria to break down cellulose, but pigs do not, even though they also consume plant material.

CONCLUSION

5 Summarise the key features of the digestive systems of mammals and how this relates to diet.

RATING MY LEARNING	My understanding improved	Not confident ◄—► Very confident ○ ○ ○ ○ ○	I answered questions without help	Not confident ◄—► Very confident ○ ○ ○ ○ ○	I corrected my errors without help	Not confident ◄—► Very confident ○ ○ ○ ○ ○

 ISBN 978 1 4886 1931 1

PRACTICAL ACTIVITY 2.3

Tubes for transport—vascular tissue in plants

Suggested duration: 90 minutes

INTRODUCTION

Tissues that are specialised for transporting substances to and from cells in plants are called vascular tissues. One tissue transports water and inorganic nutrients upwards through the plant and is called xylem. The other tissue transports sugars (in solution) produced by photosynthesis and other manufactured products throughout the plant and is called phloem.

In stems, xylem and phloem tissue form vascular bundles, with the phloem on the outer surface of the bundle. A layer of cells called the cambium (that produces secondary tissues) runs through the vascular bundle separating the xylem and phloem.

Vascular tissue is easily visible in leaves. The parallel veins of grasses and the branching veins in most other leaves are part of the vascular network of the plant. Vascular tissue extends from the roots to the very tips of leaves, and into developing buds and fruit.

MATERIALS

- stick of celery that has been standing in a solution of water and food dye
- stick of celery that has been standing in water without dye
- iodine stain
- stereoscopic microscope
- light microscope
- scalpel
- 2 × dissecting needles
- forceps
- microscope slides and coverslips
- single-edged razor blade (new or in very good condition)
- dissecting board
- paper towel and paper tissues
- prepared slides of transverse and longitudinal sections of a sunflower (*Helianthus*) stem

PURPOSE

To investigate the pathway of water movement in the stem and leaves of a flowering plant.

To examine the microscopic structure of the tissue through which water moves in a plant.

To become familiar with the structure and function of a vascular bundle.

PROCEDURE

1. Collect a celery petiole (stalk that attaches leaf to stem) that has been standing in coloured dye solution. Rinse the dye from the end and examine the petiole and leaf for evidence of dye distribution. Holding it up to the light may be helpful.
2. Place the petiole on a dissecting board and, using the razor blade, cut transverse sections (across stalk) as thinly as possible (1–2 mm thickness) in the positions shown in the diagram at right. Be sure to slice the petiole in a downward direction onto the cutting board. Arrange the sections on a microscope slide. A coverslip is not needed. Put the slide under a stereoscopic microscope and examine the cut surface of each section for the presence of dye.

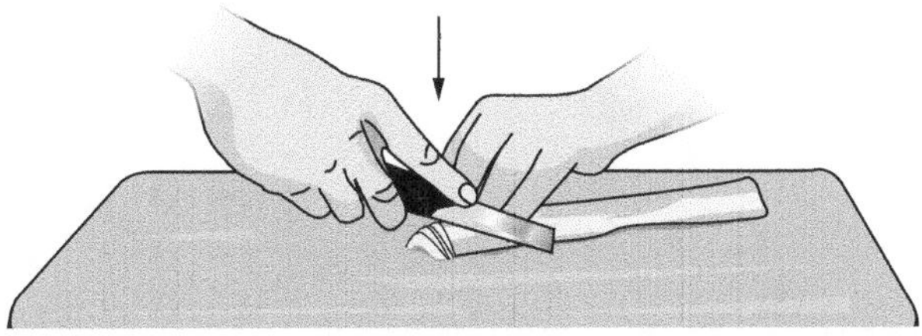

Technique for cutting sections

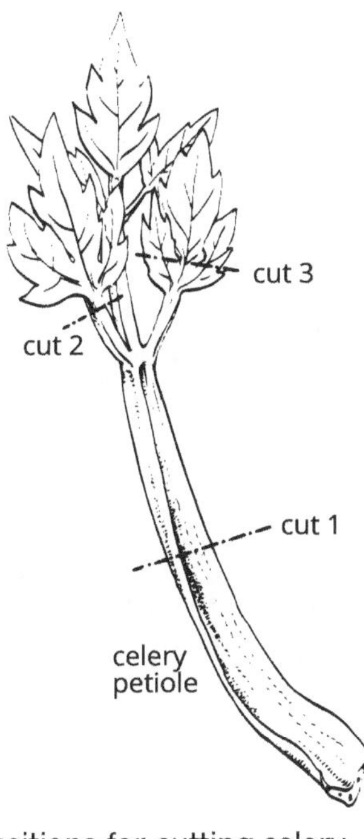

Positions for cutting celery

PROCESSING DATA

1. Describe the distribution of the dye.

2 Draw diagrams of the three sections, showing the distribution of the dye.

PROCEDURE

3 Cut a section of the petiole 1 cm thick. Stand the section on its end and cut lengthways down along one of the coloured areas. Examine this section under the stereoscopic microscope.

ANALYSIS OF RESULTS

3 Is the dye found only in definite structures within the stem? Explain.

4 Combining your observations of the transverse and longitudinal sections, draw a three-dimensional diagram of the petiole, showing where the dye has moved.

5 Does the distribution of dye necessarily show where water has moved in the stem? Explain.

 ISBN 978 1 4886 1931 1

PROCEDURE

4 Collect a 3-cm long piece of undyed celery petiole. Using the technique shown in the figure above, cut several very thin cross-sections. Put the sections on a microscope slide. Select the one that looks the thinnest and mount it on another slide in a drop of iodine stain. Add a coverslip and observe under a high power microscope using the low power objective lens. Look for some thick-walled cells in about the same position as you found the dye. These are xylem vessel cells and they form part of a vascular bundle. Water moves in xylem.

5 Cut a 1-cm thick slice from the remaining undyed celery. Place it under a stereoscopic microscope and, using a scalpel, cut down through one of the vascular bundles. Use the scalpel and dissecting needles to carefully dissect out a length of the bundle. Make sure that you take only a small piece and that there are few if any cells from either side of the bundle. Transfer the bundle to a drop of iodine on a microscope slide. Thoroughly tease it apart with the needles. Place the slide on a piece of paper towel and add a coverslip. Move the coverslip from side-to-side while applying gentle pressure with your finger, then fold the paper tissue over the coverslip and push down on it firmly with your thumb to squash the bundle. Add some more stain at the edge of the coverslip if too much has been removed.

6 Observe the slide, first using the low-power objective lens, then carefully changing to the high-power lens. Identify the cell types shown in (a) and (b).

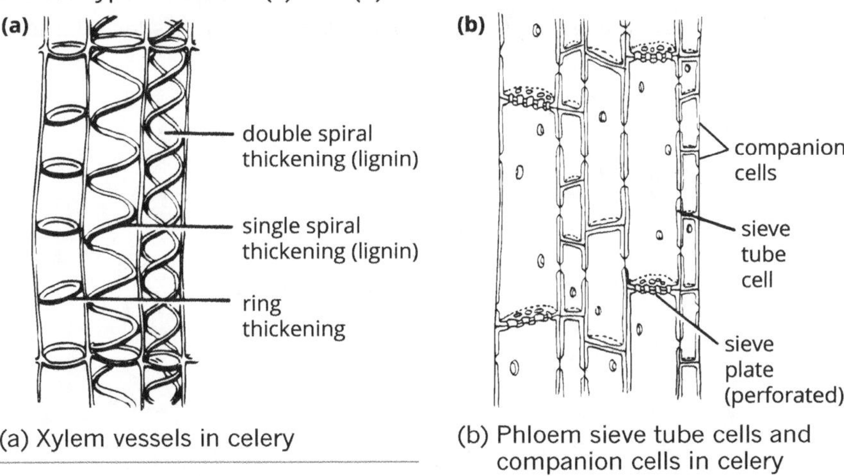

(a) Xylem vessels in celery

(b) Phloem sieve tube cells and companion cells in celery

PROCESSING DATA

6 Make drawings of several cells of each type as observed under high power magnification.

ANALYSIS OF RESULTS

7 The walls of xylem vessel cells are thickened by a material called lignin. You may also observe spiral coils in the vessel cells. These too are made of lignin. What advantage would both of these features of xylem have for a plant?

PROCEDURE

7 Collect the two prepared slides of sunflower (*Helianthus*) stem. One is a transverse section, the other is a longitudinal section. In these slides, the thickening material, lignin, is stained red.

8 Examine the transverse section. You will see a number of vascular bundles. Within a vascular bundle, you will see two distinctly different tissues—phloem and xylem. Complete Processing data 8.

9 Examine the longitudinal section. Locate a vascular bundle and again identify the xylem and phloem. Xylem vessels in *Helianthus* will show the same spiral thickening that you observed in celery. Move the slide so that you follow up or down a column of xylem tissue until you find a part where the ends of two vessels meet. Complete Processing data 9.

10 Phloem is made up mostly of sieve tube cells. Move the slide so that you follow up or down the phloem until you find a part where two sieve tube cells join. You may be able to see the sieve plate between them. You may also observe smaller cells in close association with the sieve tube cells. These are likely to be companion cells. Complete Processing data 10.

PROCESSING DATA

8 Draw a labelled diagram of a vascular bundle showing the arrangement of phloem and xylem tissue. Refer to resources if necessary.

9 Draw several xylem vessel cells, showing how they are connected vertically to each other.

ISBN 978 1 4886 1931 1

PRACTICAL ACTIVITY 2.3

10 Draw several sieve tube cells, showing how they are connected vertically to each other. Include companion cells in your drawing if you see any.

CONCLUSION

11 Using the observations made in this activity, prepare a labelled diagram showing the structure of a vascular bundle in a flowering plant. Describe the role of the following types of cells:

- xylem vessel
- sieve tube cells
- companion cells.

RATING MY LEARNING	My understanding improved	Not confident ◄——► Very confident ○ ○ ○ ○ ○	I answered questions without help	Not confident ◄——► Very confident ○ ○ ○ ○ ○	I corrected my errors without help	Not confident ◄——► Very confident ○ ○ ○ ○ ○

PRACTICAL ACTIVITY 2.4

The heart of the matter—heart structure and function

Suggested duration: 50 minutes

INTRODUCTION

The human heart first begins to beat about three and a half weeks after conception, when the fetus is only a few millimetres long. It continues to beat for the rest of life. This is an enormous amount of biological work. In an adult, on average, five litres of blood is pumped through the heart each minute and this can increase five-fold if the demands of the body require it. On average, all of the blood in the body passes through the heart in less than one minute.

MATERIALS

- sheep heart
- dissecting board
- newspaper
- scalpel
- scissors
- blunt probe
- disposable gloves
- disposal bag
- disinfectant
- antiseptic hand wash

PURPOSE

To investigate the structure of a mammalian heart and how its features make it well suited to its role.

- Biological material is potentially hazardous and should be handled and disposed of with care.
- Gloves and lab coat should be worn.
- Caution should be exercised when using the sharp dissecting tools. Take care to point the instruments away from yourself when cutting.

PROCEDURE

1 Place the heart on the dissecting board in the orientation shown in the instructional diagram on the following page—the right (thin-walled) ventricle should be on your left. Identify as many of the surface features as you can.

Use a blunt probe to gently investigate where each of the vessels leads to or comes from. Do not use force on the probe, to avoid damaging the tissue.

Features to look for (refer to the diagrams of the heart on the following page as a guide):

- Right and left ventricles separated by a muscular wall—the ventricular septum. The base of the left ventricle forms the heart apex. Feel the thickness of the muscles in the ventricle walls.
- Right and left atria.
- The 'groove' between the right atrium and ventricle and the left atrium and ventricle is called the atrio-ventricular groove. It is usually filled with white fatty tissue, so is not always immediately recognisable as a groove.
- Venae cavae (veins)—two thin-walled vessels that enter the right atrium. Look carefully on the dorsal side of the atrium to locate these. They may be collapsed, so use the probe or your fingers. Note that this may not be visible in some hearts, where they have been cut off.
- Pulmonary artery—thick-walled vessel that leaves the right ventricle.
- Pulmonary vein—thin-walled vessels that enter the left atrium. Note that these may have collapsed or may have been cut off.
- Aorta—a very thick-walled vessel that leaves the left ventricle.

ISBN 978 1 4886 1931 1

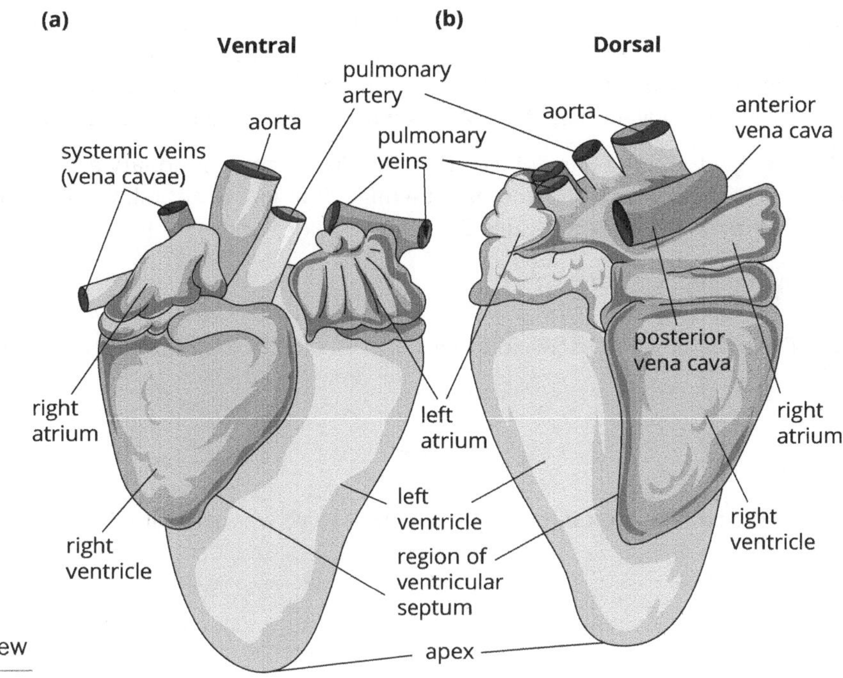

(a) Heart—ventral view (b) Heart—dorsal view

ANALYSIS OF RESULTS

1 How does the thickness and elasticity of the walls of the veins entering the atria compare with that of the arteries leaving the ventricles?

PROCEDURE

Examination of the right side of the heart

2 Use a scalpel to make cut 1, as shown in the instructional diagram below. Remember to work with the scalpel pointing away from you. Cut through the ventricle wall, along the pulmonary artery and then along the upper side of the atrio-ventricular groove. Observe the thickness of the right ventricle wall. Carefully open out the ventricle wall.

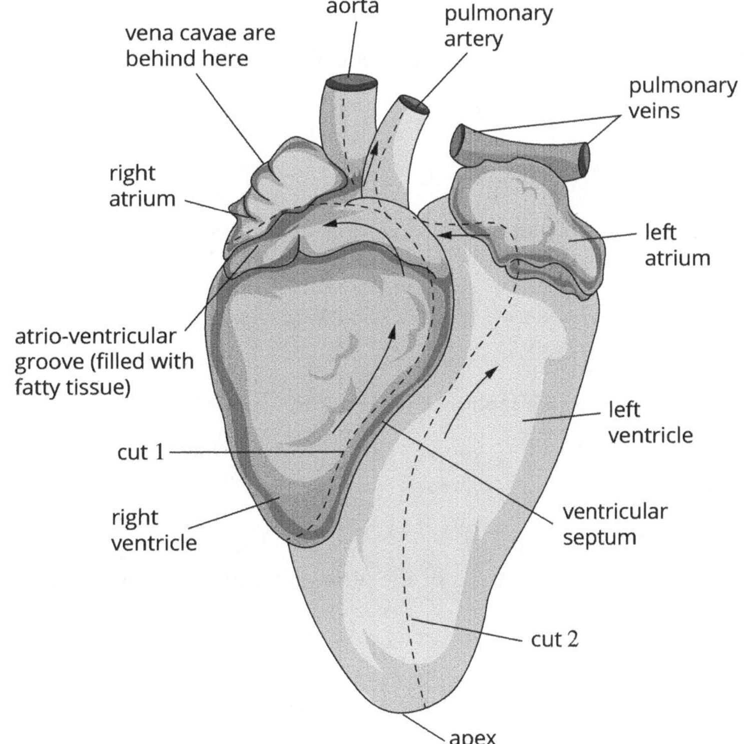

Instructional diagram: cutting sequence for dissection

3 Identify the flaps of tissue attached by thread-like tendons to the ventricle wall. These form the right atrio-ventricular (tricuspid) valve. Use the probe to gently investigate the way the flaps are arranged. Think about how these might operate. Try pushing them from the atrium side and then do the same from the ventricle side. These valves are sometimes called parachute valves, which may give a clue about how they work. Complete Analysis of results 2.

4 Identify the three flaps of tissue at the base of the pulmonary artery. These form the pulmonary semi-lunar valve. This tissue is very thin and may be difficult to locate at first. Gently slide the probe down the wall of the artery. Complete Analysis of results 3.

ANALYSIS OF RESULTS

2 Describe how the tricuspid valve works to direct blood flow between the right atrium and right ventricle.

3 Describe how the pulmonary semi-lunar valve works to direct blood flow at this point.

PROCEDURE

Examination of the left side of the heart

5 Use the scalpel to open the left ventricle by cutting from the apex along cut line 2 (see instructional diagram on the previous page). Ensure the scalpel blade is pointing away from you. Note the thickness of the ventricle wall compared to the right ventricle wall. Look for the atrio-ventricular (bicuspid/mitral) valve inside the left side of the heart. Continue cutting until you reach the base of the aorta, at the point where it leaves the ventricle. Use the probe to locate the aortic semi-lunar valve at the base of this artery.

6 Just above the semi-lunar valves you should be able to locate two small openings where the coronary arteries branch from the aorta. Use the probe to gently explore where these arteries lead. Complete Analysis of results 4.

7 At the conclusion of the dissection, wrap the heart in the newspaper and place in the disposal bag supplied for this purpose. Follow your teacher's instructions to manage soiled utensils and wipe down benches. Dispose of gloves and wash hands thoroughly using antiseptic handwash.

ANALYSIS OF RESULTS

4 Describe how the bicuspid valve works to direct blood flow.

5 Describe how the aortic semi-lunar valves work to direct blood flow.

6 Where do the coronary arteries lead and what is their role?

7 a Compare the thickness of the walls of the atria and ventricles. Account for this difference.

 ISBN 978 1 4886 1931 1

b Compare the thickness of the walls of the right ventricle and left ventricle. Account for this difference.

8 Look carefully at the diagram of the heart shown below.

a Clearly label the different parts of the heart, selecting from the terms listed below.

aorta	pulmonary artery	semi-lunar valves	right atrium
left ventricle	atrio-ventricular (tricuspid) valve	pulmonary vein	venae cavae
left atrium	septum	right ventricle	atrio-ventricular (bicuspid) valve

b Use red and blue coloured pencils to shade the two sides of the heart, using red to show oxygenated blood and blue to show deoxygenated blood. Add a legend to the diagram to illustrate the meaning of the colours used.

c Add arrows to clearly show the direction of blood flow into, through and out of the heart. Indicate where blood is coming from and where it is going to.

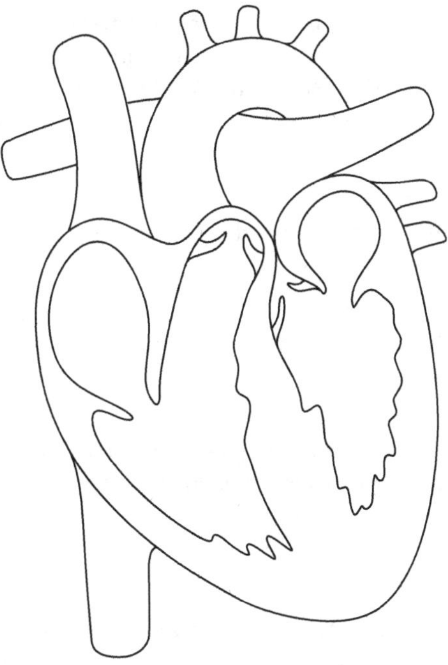

CONCLUSION

9 Suggest why the mammalian heart is described as a double pump.

10 Summarise the role of the heart, outlining how at least two different features make it well suited to this role.

RATING MY LEARNING	My understanding improved	Not confident ⟷ Very confident ○ ○ ○ ○ ○	I answered questions without help	Not confident ⟷ Very confident ○ ○ ○ ○ ○	I corrected my errors without help	Not confident ⟷ Very confident ○ ○ ○ ○ ○

PRACTICAL ACTIVITY 2.5

Rat dissection—mammalian systems

Suggested duration: 90 minutes

INTRODUCTION

Cells are the building blocks of multicellular organisms. Specific cell types are organised into tissues, tissues are organised into organs, and organs into systems. Each system is composed of a number of organs and components that work together to serve a particular role within the whole organism.

Dissecting a rat provides an opportunity to examine several mammalian body systems. Of particular interest are the circulatory, digestive and respiratory systems. Despite the difference in size, the basic structure and organisation of these systems is similar in both rats and humans.

While there are instructions to guide you through this investigation, you are encouraged to actively pursue further questions and considerations as you proceed. Take the opportunity to write notes and questions that can form the basis of further investigation and discussion. Remember that record taking can be in many forms, including photographs. Think about how the different organs and systems examined in this dissection are interconnected to ensure the overall functioning of the organism.

MATERIALS

- rat pinned out on dissecting board
- scalpel
- dissecting scissors
- forceps
- probe
- dissecting pins
- newspaper
- ruler
- disposable gloves
- disposal bag
- disinfectant
- antiseptic hand wash

PURPOSE

To dissect a rat and compare the structure and organisation of the major organs and body systems to that of a human.

- Biological material is potentially hazardous and should be handled and disposed of with care.
- Gloves and lab coat should be worn.
- Caution should be exercised when using the sharp dissecting tools. Take care to point the instruments away from yourself when cutting.

PROCEDURE

1 Examine the mouth and teeth of the rat. It may be necessary to gently prise the mouth open for a better view. Consider what the arrangement of teeth suggests about the diet of the rat and how this compares to humans.

2 Follow your teacher's direction (and demonstration) to open the abdomen and chest cavity of the rat. Making the initial incision is often the most difficult. At this point, it is important not to damage the organs underneath.

3 Once the skin has been peeled and pinned back, using either a photo or diagram (depending on your teacher's directions) record the initial position of each of the organs. Consider where each organ sits within the chest and abdominal cavities and the other organs nearby.

4 Identify and classify the organs associated with each of the systems you have studied—circulatory, digestive and respiratory systems. Make five brief observations at this point of the dissection. Note if there is anything unexpected or surprising about the arrangement or position of organs.

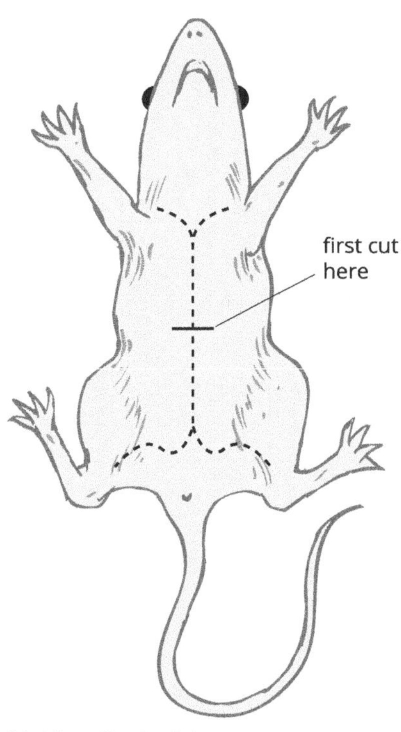

Making the incision

ISBN 978 1 4886 1931 1

5 The stomach and pancreas can be viewed by gently lifting the liver with the probe.

6 Follow the pathway of the stomach to the small intestines, then large intestines. Take care not to puncture the intestines as you do so.

7 The intestinal tract can be removed by gently freeing the mesentery (membranous tissue) holding the intestines together. The mesentery contains arteries and veins that supply blood to the intestines.

8 With the intestinal tract removed, identify the caecum (a bag like structure that signals the beginning of the large intestine).

9 Identify the colon next, and finally the rectum, which will contain pellet-shaped faeces.

10 Compare the length of the small intestine to the large intestine. Record the length of each using a ruler.

11 Identify the location of each of the kidneys. Note their relative size.

12 Under your teacher's directions, cut through the rib cage to expose the heart, lungs and diaphragm.

13 Upon completion of the dissection, ensure that you follow directions for the appropriate disposal of the rat and clean-up procedures.

PROCESSING DATA

1 Use the diagram of the rat provided to prepare a detailed and labelled sketch of the organs and components that were observed during the dissection.

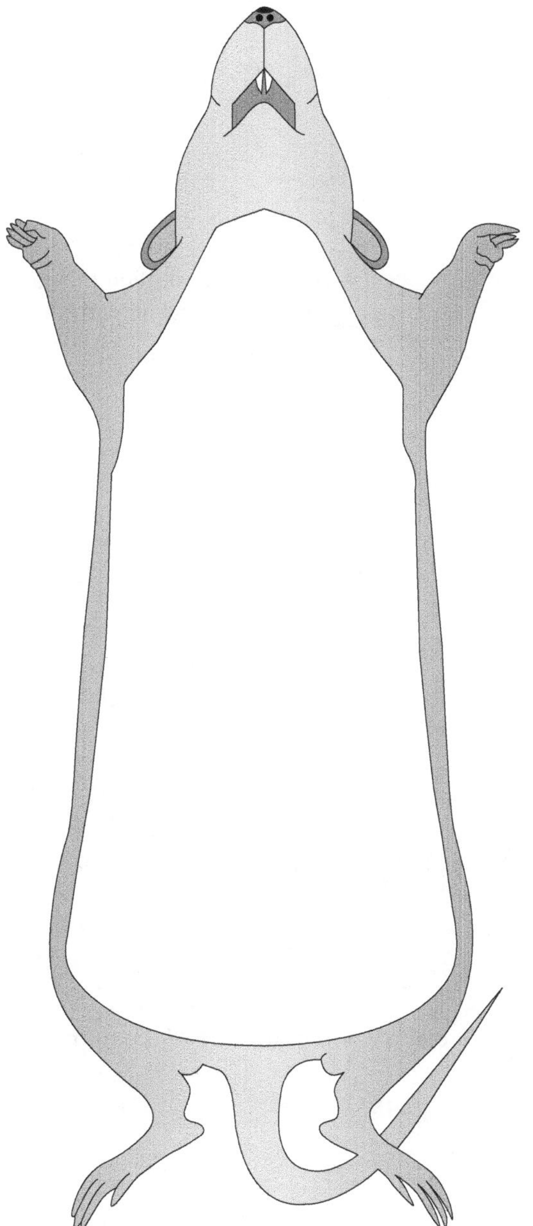

ISBN 978 1 4886 1931 1

2 Complete Table 1 to summarise your findings.

TABLE 1 Summary of findings during dissection of rat

Body system	Organs/structures	Key observations

ANALYSIS OF RESULTS

3 For each of the organs/structures you have listed in Table 1, outline the role it performs and describe how the features of each organ/structure make it well suited to that particular role.

4 What can you determine about the diet of a rat from its teeth?

5 Compare and contrast the teeth and diet of the rat to those of the human.

 ISBN 978 1 4886 1931 1

6 Compare the length and width of the rat's small intestine to its large intestine. How does the length of the small intestine assist its role?

7 Investigate how the digestive system of a typical carnivore compares to that of a typical omnivore or herbivore. Provide an example of each in your response.

8 Humans are omnivores. How is the human digestive system different from that of herbivores or carnivores? Describe specific examples.

CONCLUSION

9 Describe how the tissues and organs of the digestive system are organised into a functioning system.

10 Summarise the relationship between the structure and function of the digestive system. Include two specific examples to illustrate your response.

RATING MY LEARNING	My understanding improved	Not confident ◄—► Very confident ○ ○ ○ ○ ○	I answered questions without help	Not confident ◄—► Very confident ○ ○ ○ ○ ○	I corrected my errors without help	Not confident ◄—► Very confident ○ ○ ○ ○ ○

PRACTICAL ACTIVITY 2.6

Investigating and modelling biological structures—an alternative to dissection

Suggested duration: 90 minutes

INTRODUCTION

While animal dissections can provide insights into the structure of organs and systems, there are many alternative approaches. Biological structures can also be investigated using physical and digital models such as interactive anatomy applications, virtual dissections, videos or 3D models. These alternative approaches reduce the use of animals in the classroom and some even allow interaction with working structures and systems (for example, a simulation of a beating heart pumping blood through the circulatory system).

MATERIALS

- Internet access
- paper
- pencil
- materials to construct model

PURPOSE

To construct a model of an organ, tissue or cell from the circulatory or respiratory system, enhancing your understanding of its structure, function and complexity.

PROCEDURE

1 With guidance from your teacher, find resources to investigate the circulatory or respiratory system. Resources may be in the classroom (for example, plastic models) or online (virtual dissections, interactive anatomy applications or videos). Most resources focus on human (mammalian) systems but you may choose to investigate the systems of fish, amphibians, reptiles or birds.

2 Explore the system's structures and their arrangements. Think about the role each structure plays in the functioning of the system. Make notes of any important features you notice.

3 Choose a structure (organ, tissue or cell) within this system of which you will construct a model. You can do this individually or as part of a group.

4 Plan how you will construct your model, thinking about the materials that are available to you and the format that will best represent your chosen organ, tissue or cell. Some ideas are modelling clay, paper, plastic bottles or digital drawing tools. Sketch a draft of your model, labelling structures and making notes of the materials you will use. Make sure you include each of the important internal and external structures.

5 Construct your model using your chosen materials.

DISCUSSION

1 Describe how the cells, tissues and organs of your chosen system are organised into a functioning system. Draw a labelled diagram to support your answer.

ISBN 978 1 4886 1931 1

2 How does the circulatory system and its structures support life?

3 Describe the role of your chosen organ, tissue or cell in the system.

4 Identify at least three structures that are important to the functioning of your chosen organ, tissue or cell and describe their roles.

5 Reflect on your experience during this activity and discuss its advantages and disadvantages as an alternative to animal dissection.

RATING MY LEARNING	My understanding improved	Not confident ◄——► Very confident ○ ○ ○ ○ ○	I answered questions without help	Not confident ◄——► Very confident ○ ○ ○ ○ ○	I corrected my errors without help	Not confident ◄——► Very confident ○ ○ ○ ○ ○

DEPTH STUDY 2.1

Restricting the flow—water regulation versus carbon dioxide exchange in plants

Suggested duration: Part A—90 minutes; Part B—45 minutes; Part C—90 minutes

INTRODUCTION

The chlorophyll-containing mesophyll cells of leaves need a supply of water and carbon dioxide in order to photosynthesise. Carbon dioxide enters the leaf from the external environment through stomata, while water enters the plant at the roots and is drawn upwards through the xylem to the leaves. As a consequence, when the stomata are open to take in carbon dioxide, water vapour can escape from the intercellular spaces inside the leaf to the external environment. Leaves cannot afford to lose too much water in this way, especially if they live in dry environments where water availability may be limited. A feature of many plant species to reduce water loss in this way is an uneven distribution of stomata on the upper and lower surface of leaves.

In this primary investigation you will investigate the potential for water loss from upper and lower surfaces of leaves and then take steps to verify your experimental outcomes. Finally, you will plan and conduct an investigation to establish whether or not a pattern exists in the distribution of stomata on the leaves of plants.

MATERIALS

For each group of four students:

- 4 × measuring cylinders (either 20 mL or 50 mL)
- petroleum jelly on a watch glass
- newspaper
- paper towel
- paraffin oil
- electronic balance (accurate to 0.01 g)
- 4 × plant cuttings from the same species
- 4 × cardboard disks with central cross cut (to support cuttings in measuring cylinder)

PURPOSE

To investigate water loss from upper and lower leaf surfaces.

To consider the effect of water conservation on the supply of carbon dioxide.

To consider whether an uneven distribution of stomata on upper and lower leaf surfaces is a common feature in plants.

Part A—investigating water loss

CONDUCTING YOUR INVESTIGATION

1. Examine the orientation of the leaves for the plant specimen you will be using. Do both upper and lower surfaces have equal exposure to sunlight? What difference might this make to water conservation by the plant? Make notes as you go as this will help you answer Questioning and predicting 1 and 2.
2. Collect four measuring cylinders and label them 1–4.
3. Fill each cylinder with water to the top graduation.
4. Collect four plant cuttings—there will be one to place in each measuring cylinder. Make sure that each cutting has the same number of leaves of roughly the same size. Do not leave cut stems out of water for too long.
5. This experiment requires four different treatments to the leaves of plant cuttings:
   - Cutting 1: untreated—no petroleum jelly on leaves
   - Cutting 2: upper epidermis coated—smear a thin layer of petroleum jelly on the upper surface of each leaf
   - Cutting 3: lower epidermis coated—smear a thin layer of petroleum jelly on the lower surface of each leaf
   - Cutting 4: upper and lower epidermis coated—smear a thin layer of petroleum jelly on both the upper and lower surface of each leaf.

   Prepare cuttings 1–4 as described above. When applying petroleum jelly to only one surface of a leaf, it is important to cover the entire surface and to avoid contact of the untreated epidermis with the petroleum jelly. This is easily achieved by treating each leaf on a fresh area of the newspaper. When you have finished, dispose of the newspaper and wipe your hands.

ISBN 978 1 4886 1931 1

6 Push the stem of each cutting through the central cross in the cardboard disk. Place the stem and disk in the measuring cylinder, immersing the cut stem in water. The disk should support the cutting centrally so that it does not tip over. Label each measuring cylinder 'Cuttings 1–4' to make them easily identifiable.

7 Carefully pour paraffin oil down the inside of each measuring cylinder until it forms a visible layer on the water surface.

8 Weigh each measuring cylinder to the nearest 0.01 g. Record the data in Table 1. Set your group's cuttings aside in a place designated by your teacher.

TABLE 1 Cutting weights and percentage water loss

Leaf treatment	Initial mass ($mass_I$) g	Final mass ($mass_F$) g	Percentage water loss $\frac{(mass_I)\,g - (mass_F)\,g}{(mass_I)\,g} \times \frac{100}{1}$	Percentage water loss (class average)
untreated				
upper epidermis coated				
lower epidermis coated				
upper and lower epidermis coated				

24–36 hours later

9 Reweigh each measuring cylinder, again to the nearest 0.01 g. Record the weights in Table 1.

10 Dispose of your plant and measuring cylinder as directed by your teacher.

QUESTIONING AND PREDICTING

1 Develop a question about leaves, stomata and water loss for the plant specimen you are focusing on in this investigation.

2 State a hypothesis relevant to this investigation.

PROCESSING DATA AND INFORMATION

3 Calculate the percentage water loss for your cutting. Enter the data into Table 1.

4 Collect the class percentage water loss data for each of the different leaf treatments. Enter the data into Table 1.

ANALYSING DATA AND INFORMATION

5 a Which leaf treatment showed the least percentage water loss?

b How does this compare with the value for the untreated leaves?

6 Which leaf treatment showed the greatest percentage water loss? Explain.

7 Outline the purpose of the cutting with the untreated leaves.

8 Explain why a layer of paraffin oil was applied to the surface of the water.

9 Were any of the percentage water loss figures very similar? If so, suggest an explanation to account for this.

10 From the experimental data, what inferences can be made about the distribution of stomata on the leaves of this plant?

Part B—checking your inferences

CONDUCTING YOUR INVESTIGATION

1 Use either the 'epidermal peel' or 'nail polish' method, as demonstrated by your teacher, to make an 'impression' of both upper and lower leaf surfaces of your plant.

2 Choose a magnification (preferably between ×50 and ×100) that enables you to clearly see stomata if they are present.

3 Mount the upper epidermal peel first. Count the number of stomata in a field of view, record the number, move the slide across to another area and repeat. Do this two more times then calculate the average number of stomata per field of view. Repeat the process at the same magnification for the lower epidermal peel.

4 Record the average number of stomata per field of view in Table 2.

MATERIALS

For each group of students:

- materials for making epidermal peels (nail polish and sticky tape)
- microscope slides and coverslips
- light microscope

TABLE 2 Stomatal count

Epidermis	Average number of stomata per field of view
upper	
lower	

COMMUNICATING

1 Do your observations support the inferences you made about stomatal distribution in Part A? Explain.

 ISBN 978 1 4886 1931 1

2 Is your hypothesis supported or not supported? Outline the experimental evidence that supports your claim.

3 Explain the significance of the distribution of stomata.

4 Comment on the distribution of stomata you would expect to observe in species adapted to dry conditions. Explain your answer.

5 Predict and explain the effect of reducing water loss on a plant's:

a ability to take up carbon dioxide from the atmosphere

b rate of photosynthesis

6 Which of the two factors—water conservation or carbon dioxide uptake—do you think would be most critical in determining the distribution and density of stomata? Give reasons for your answer.

Part C—comparing cuttings

In this part of the activity you will plan and conduct your own investigation. This time the focus is on more than one kind of plant. You will be investigating whether or not the pattern observed for the plant specimen in Parts A and B is common to plants in general.

You will have to think about a logical approach to this investigation and select materials that will be useful in conducting the laboratory work. You will collect, process and analyse the experimental data and evaluate the questions and hypothesis. It will be useful to refer to the Practical investigation notes of your Biology toolkit for assistance. GO TO ➤ pages ix–xiv

PURPOSE

To investigate and compare the patterns of stomatal distribution on upper and lower leaf surfaces of different species of plants.

PLANNING YOUR INVESTIGATION

Use your knowledge and experience of working in an experimental setting to design an investigation that will meet the purpose set out below. Think about a logical approach to acquiring the information you seek. Remember that scientific investigations must be unbiased and designed to provide reliable data. If you are working with several plant species, collaboration with other students or groups of students may be an efficient approach. Think carefully about the different kinds of plants you will investigate. You may decide to include a range of plants from similar habitats and environmental conditions, or plants from different habitats and environmental conditions, for example, hot dry climates, moderate climates and cool moist climates.

1 Write a list of materials you will need to conduct this investigation.

Materials

-
-
-
-
-
-
-
-

2 Think about any risks associated with the listed materials. Identify the risks and outline safety measures that will be important to address these.

CONDUCTING YOUR INVESTIGATION

3 Prepare a logical sequence of step-by-step instructions that another student could easily follow. Check with your teacher before proceeding with the activity.

 ISBN 978 1 4886 1931 1

QUESTIONING AND PREDICTING

4 Develop a question about leaves, stomata and water loss for the plant specimens in this investigation.

5 State a hypothesis relevant to this investigation.

PROCESSING DATA AND INFORMATION

6 Think about a logical way to record data from the investigation. Remember that tables are useful for displaying information succinctly and provide a convenient means of comparing data.

ANALYSING DATA AND INFORMATION

7 Examine your tabulated data. Explain what the information means—what does it tell you about the distribution of stomata on the upper and lower surfaces of leaves from different kinds of plants?

CONCLUSION

8 Evaluate your hypothesis in this investigation. Was it supported or not supported? Outline the experimental evidence that supports your claim.

MODULE 2 • REVIEW QUESTIONS

Multiple choice

1 The cells of complex multicellular organisms are organised in ways that facilitate efficient overall functioning for the organism. This organisational hierarchy can be summarised by the following order:

A cells, organs, tissues, systems, organism

B cells, tissues, organs, organism, systems

C cells, tissues, organs, systems, organism

D tissues, organs, systems, cells, organism

2 The cell shown is a specialised plant cell. The features of this cell suggest it is likely to be involved in:

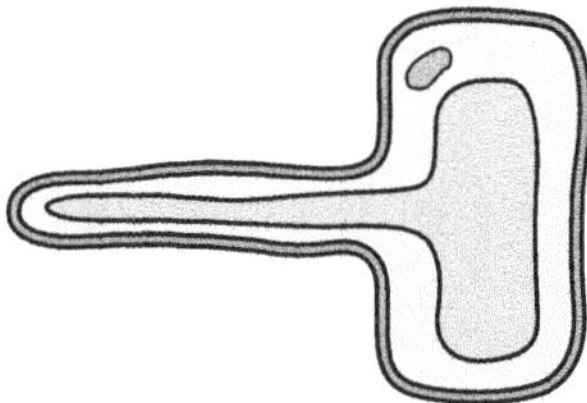

A photosynthesis

B absorption

C reproduction

D transpiration

3 Which of the following statements in relation to autotrophs and heterotrophs is not correct?

A Autotrophs are organisms that manufacture their own organic compounds from inorganic materials obtained from the environment in a process called chemosynthesis.

B Heterotrophs are organisms that are unable to manufacture their own organic compounds and so rely on consuming other organisms to acquire these.

C Autotrophs are organisms that manufacture their own organic compounds from inorganic materials obtained from the environment through processes such as photosynthesis or chemosynthesis.

D Heterotrophs are represented by organisms such as animals and fungi.

4 Complex carbohydrates need to be digested before cells can access them for use in cellular respiration and the production of energy. Cellulose represents one example of complex carbohydrates found in the cell walls of plant material. Cellulose digestion:

A relies on the presence of fermentation bacteria in herbivores

B occurs at the junction of the small and large intestines in herbivores and omnivores

C in herbivores, is characterised by the presence of a rumen

D occurs in the caecum of herbivores, carnivores and omnivores

5 Large multicellular plants are characterised by the presence of specialised vascular tissue involved in the transport of materials throughout the plant. This includes:

A the translocation of water up and down the plant through the xylem

B the transport of organic compounds produced in photosynthesis in a process called transpiration

C the movement of inorganic materials produced in photosynthesis through the phloem

D the movement of water in an upward direction from the roots to the shoot system in tissue called xylem

6 Select the description below that does not represent a feature of efficient gas exchange surfaces in animals.

A moist surface

B presence of alveoli

C large surface area

D rich blood supply

Short answer

7 Examine the diagrams below showing the digestive systems and tooth profiles for three different heterotrophs.

X

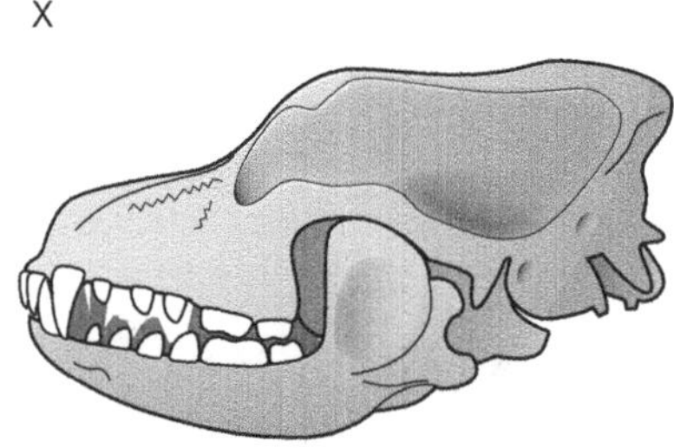

Y

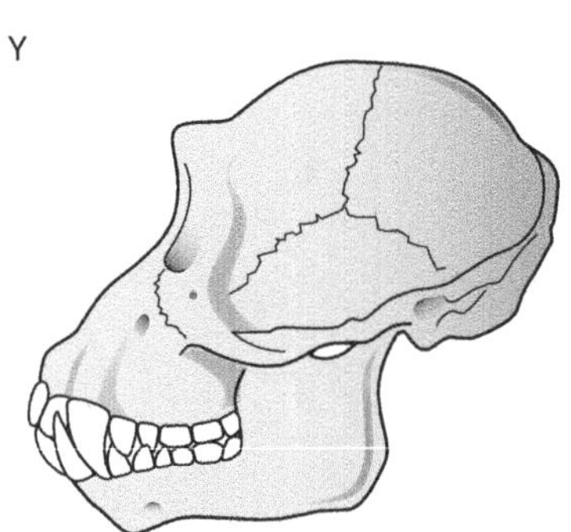

Z

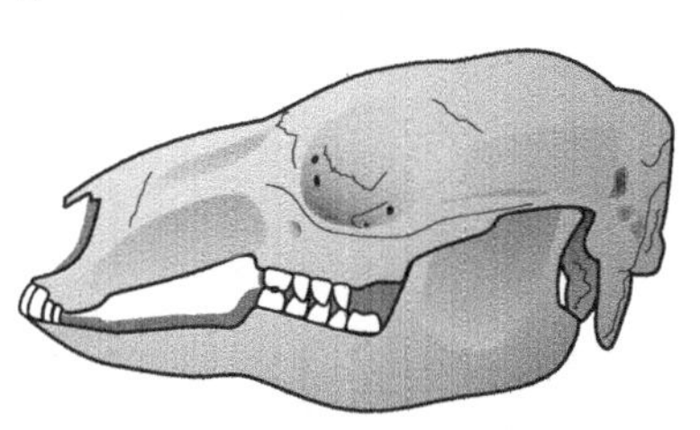

ISBN 978 1 4886 1931 1

a Select the tooth profile that corresponds to each digestive system. Record this in the second column of the table.

b Complete the table by describing the diet of each animal, explaining your reasoning in the final column.

Digestive system	Tooth profile	Diet	Explanation
A 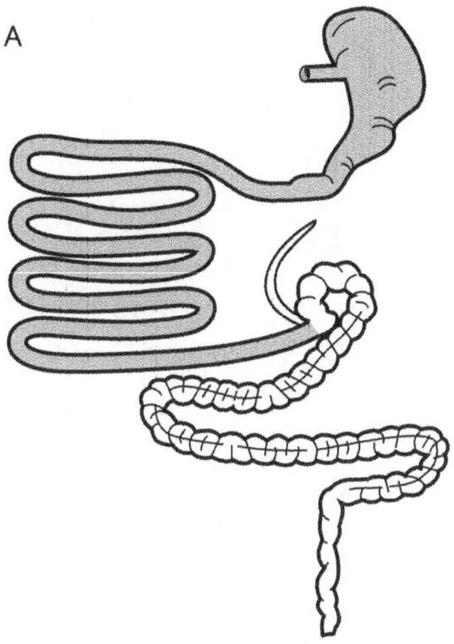			
B 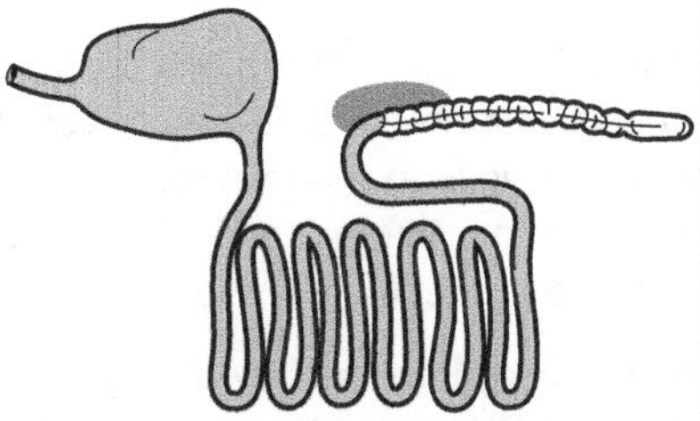			
C 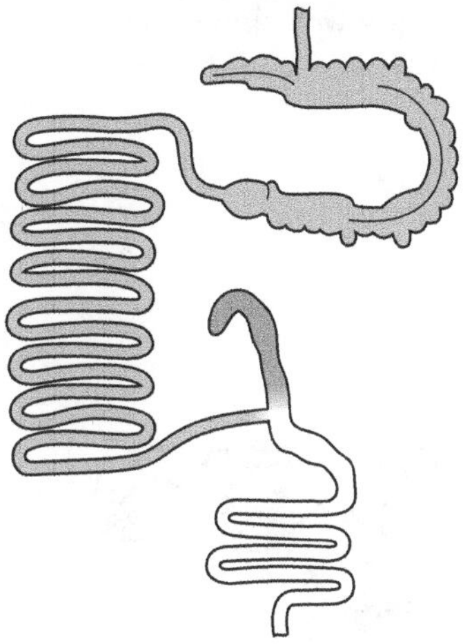			

8 The following graph summarises water loss through transpiration from a garden plant throughout the day.

(a)

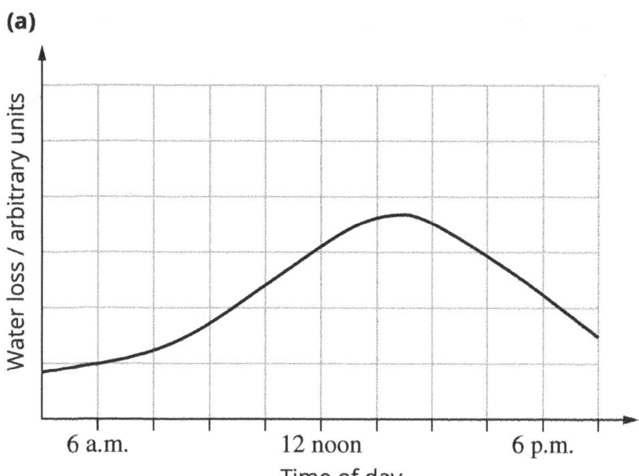

(b)

a Define transpiration.

b What time of the day was the temperature likely to have been the hottest? Explain why you think so.

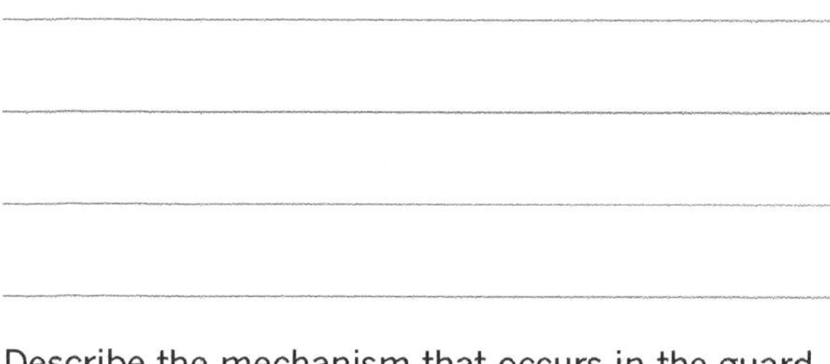

c Describe the mechanism that occurs in the guard cells as they change from state X to state Y.

d Consider the diagrams of stomata shown. Decide which state the stomata of this plant are likely to be in at the time you have identified in b. Explain why you think so.

e Explain how the stomata operate to balance the carbon dioxide (for photosynthesis) and water conservation needs of plants.

9 A group of students preparing for a test on the topic of circulatory systems found the following questions on past test papers, along with model answers to each. The questions focus on different aspects of mammalian circulation. Provide model answers to the questions.

a Capillaries are characterised by a very thin wall, only a single cell thick. Explain how this feature makes a capillary well suited to its function.

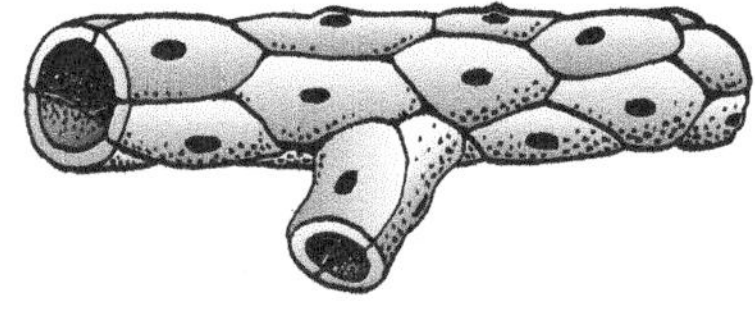

b A student looking for patterns in the way blood flows through different kinds of vessels stated that 'Deoxygenated blood always flows through veins'. Explain whether you agree or disagree with this statement.

 ISBN 978 1 4886 1931 1

c During development of the human fetus, the septum (wall between the two atria) is incomplete. Blood can move directly from the right to the left atrium. Oxygenation of fetal blood occurs at the placenta rather than the lungs. At birth, this hole—the foramen ovale—closes over and deoxygenated blood is directed from the right ventricle to the lungs. In some babies, the foramen ovale fails to fully close. Such infants are sometimes to referred to as 'blue babies' due to a characteristic bluish tinge on their lips and skin.

i Suggest an explanation for the bluish tinge these babies display.

ii What effect might this condition have on the health of such infants?

Extended response

10 A biology class studying digestion was set the task of designing a simulation to investigate physical and chemical digestion. The hypothesis under investigation was:

If chemical digestion proceeds more quickly after physical digestion has taken place, then small pieces of material will experience a faster rate of chemical digestion than large pieces of the same material.

a Given the list of materials below, describe an experiment to test this hypothesis.

Materials:

- antacid tablets
- mortar and pestle
- 2 × 50 mL beakers
- water
- stopwatch.

b Identify the:

i independent variable:

ii dependent variable:

c Describe two ways in which you would ensure this experiment is controlled.

d Describe in detail how this experiment simulates both physical and chemical digestion.

e Outline the results that would support the hypothesis.

f Explain the significance of enzymes in the digestive systems of organisms.

MODULE 3 Biological diversity

Outcomes

By the end of this module you will be able to:

- develop and evaluate questions and hypotheses for scientific investigation BIO11-1
- design and evaluate investigations in order to obtain primary and secondary data and information BIO11-2
- communicate scientific understanding using suitable language and terminology for a specific audience or purpose BIO11-7
- describe biological diversity by explaining the relationships between a range of organisms in terms of specialisation for selected habitats and evolution of species BIO11-10

Content

EFFECTS OF THE ENVIRONMENT ON ORGANISMS

INQUIRY QUESTION **How do environmental pressures promote a change in species diversity and abundance?**

By the end of this module you will be able to:

- predict the effects of selection pressures on organisms in ecosystems, including: (ACSBL026, ACSBL090) CCT L
  - biotic factors
  - abiotic factors
- investigate changes in a population of organisms due to selection pressures over time, for example: (ACSBL002, ACSBL094) S CCT ICT L N
  - cane toads in Australia
  - prickly pear distribution in Australia

ADAPTATIONS

INQUIRY QUESTION **How do adaptations increase the organism's ability to survive?**

By the end of this module you will be able to:

- conduct practical investigations, individually or in teams, or use secondary sources to examine the adaptations of organisms that increase their ability to survive in their environment, including: CCT ICT WE
  - structural adaptations
  - physiological adaptations
  - behavioural adaptations
- investigate, through secondary sources, the observations and collection of data that were obtained by Charles Darwin to support the theory of evolution by natural selection, for example: ICT L
  - finches of the Galapagos Islands
  - Australian flora and fauna

THEORY OF EVOLUTION BY NATURAL SELECTION

INQUIRY QUESTION **What is the relationship between evolution and biodiversity?**

By the end of this module you will be able to:

- explain biological diversity in terms of the theory of evolution by natural selection by examining the changes in and diversification of life since it first appeared on the Earth (ACSBL088)
- analyse how an accumulation of microevolutionary changes can drive evolutionary changes and speciation over time, for example: (ACSBL034, ACSBL093) CCT L
  - evolution of the horse
  - evolution of the platypus
- explain, using examples, how Darwin and Wallace's theory of evolution by natural selection accounts for:
  - convergent evolution
  - divergent evolution
- explain how punctuated equilibrium is different from the gradual process of natural selection

EVOLUTION—THE EVIDENCE

INQUIRY QUESTION **What is the evidence that supports the theory of evolution by natural selection?**

By the end of this module you will be able to:

- investigate, using secondary sources, evidence in support of Darwin and Wallace's theory of evolution by natural selection, including but not limited to: ICT L
  - biochemical evidence, comparative anatomy, comparative embryology and biogeography (ACSBL089) ICT L
  - techniques used to date fossils and the evidence produced ICT L
- explain modern-day examples that demonstrate evolutionary change, for example: ICT L
  - the cane toad
  - antibiotic-resistant strains of bacteria

Key knowledge

Effects of the environment on organisms

SELECTION PRESSURES

An **ecosystem** is a system composed of communities of living organisms interacting with one another and with their non-living surroundings. This includes organisms interacting with members of their own **species** and with organisms belonging to other species.

All of the factors in the environment, both living and non-living, impact on organisms in some way, influencing their chances of survival. These factors are referred to as **selection pressures**.

Selection pressures affect individual organisms and drive change within species as well as changes in the diversity and abundance of species. When a selection pressure affects one species in a community, other species with which it interacts will also be affected. When **keystone species** are adversely impacted, the consequences can be disruptive to the whole **community**.

ORGANISING THE ENVIRONMENT

The environment of an organism is made up of **biotic** and **abiotic** factors that have an effect on it at some time in its life.

Abiotic factors are the non-living factors in an organism's environment—its physical surroundings. Examples of abiotic factors are air, water, temperature, air pressure, oxygen availability, light, features of the land (such as plains or mountain ranges, sand or rock) and weather.

Biotic factors are the living factors in an organism's habitat. This includes other organisms of the same species and members of other species in the community.

The biotic factors in an ecosystem can be organised into:

- individuals (single organisms such as one animal, plant, fungus or unicellular organism)
- **populations** (groups of organisms of the same species living together in a defined geographic area)
- communities (groups of different species living in the same habitat and interacting with one another)
- **habitats** (places where organisms live at a particular time)
- keystone species (species that play a critical role in the stability of an ecosystem).

POPULATION CHANGES

Population density and **distribution** are influenced by many abiotic and biotic factors. Generally, when the resource requirements of a population are met and there are **limiting factors** acting on the population (for example, **predation** limiting prey numbers), population size can remain stable.

The size of a population can be affected by four processes:

- births or germination (also called **natality**)
- deaths (also called **mortality**)
- **immigration** (organisms moving into a population from another population)
- **emigration** (organisms moving out of a population).

Birth and immigration bring new population members and thus increase the population size. Death and emigration decrease the population size. Immigration and emigration are collectively known as **migration**. These four processes determine the rate of change in a population over time (Figure 3.1).

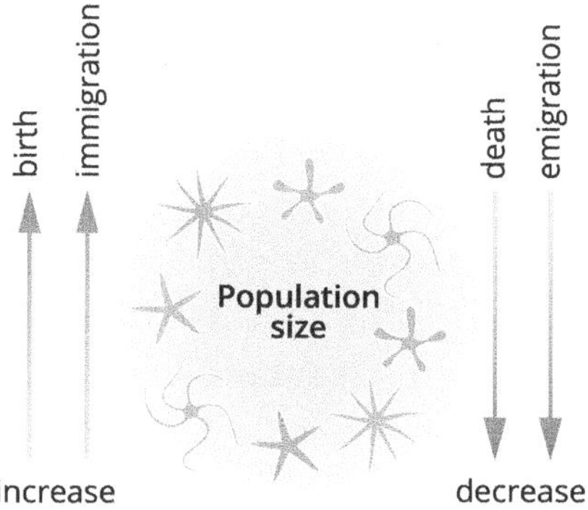

FIGURE 3.1 The size of a population depends on birth rate, death rate and migration.

Exponential population growth

When the birth rate remains higher than the mortality rate, population size increases and, depending on other factors, may eventually increase at an exponential rate (Figure 3.2a). Species that tend to experience **exponential population growth** are those that have a short generation time and give rise to large numbers of offspring.

Exponential population growth is normal for some species when environmental conditions are favourable and resources are abundant. Because these conditions generally last only a short time, exponential growth is usually short-lived. But if favourable conditions continue, then a **population explosion** may occur. The introduction of several species into Australia demonstrates this phenomenon. The prickly pear cactus, introduced with the First Fleet in 1788, is an invasive plant that is well suited to Australian conditions. It reproduced prolifically, quickly becoming a significant pest species. The introduction of the cactus moth, *Cactoblastis cactorum*, in the 1920s eventually succeeded in controlling the population. Rabbits and cane toads are examples of other introduced species that have experienced population explosions.

Ecosystem carrying capacity

In the absence of limiting factors, the population growth of a species will always be exponential. However, in a real ecosystem, population growth is affected

 ISBN 978 1 4886 1931 1

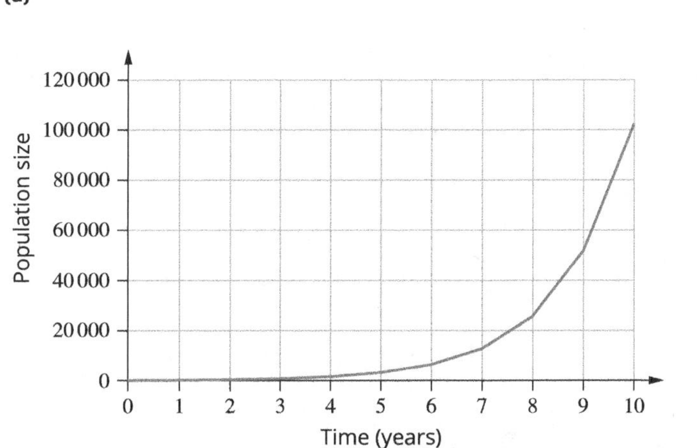

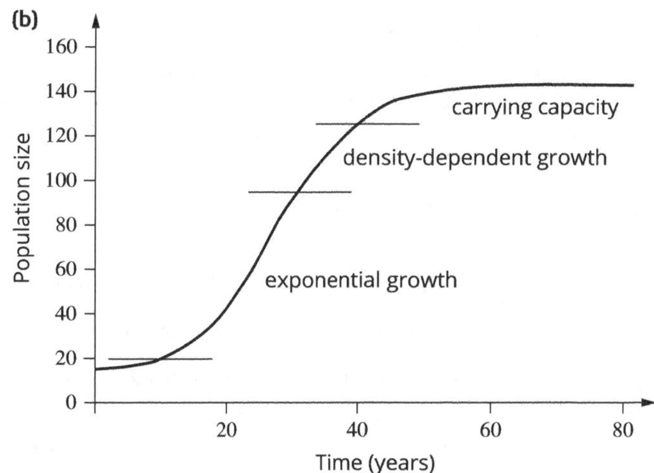

FIGURE 3.2 (a) The size of a population over time (years), showing exponential population growth. (b) Population growth curve showing exponential growth (where no limiting factors are acting), followed by density-dependent growth, where factors such as predation and competition begin to limit population growth. When population growth reaches equilibrium, the carrying capacity of the ecosystem has been reached.

by density-dependent factors such as competition for resources.

When a species' population reaches **equilibrium**, with the number of births and deaths in the population cancelling each other out, the species has reached the maximum population size that the ecosystem can sustain indefinitely. This is the ecosystem's **carrying capacity** for that species (Figure 3.2b). The carrying capacity of an ecosystem is the number of organisms of a particular species that the ecosystem can sustain.

The S-shaped graph in Figure 3.2b shows the initial exponential growth of a population, which then flattens out as it begins to be affected by density-dependent factors. The population growth rate may decline until birth and death balance each other and the population is limited to the carrying capacity of the environment.

The changing nature of ecosystems means that their carrying capacity can change over time. For example, as the availability of resources fluctuates, so too does the potential of a given ecosystem to maintain populations of organisms.

Adaptations

Adaptations are the characteristics of an organism that assist it to survive and reproduce in its environment. Adaptations are inherited.

TYPES OF ADAPTATIONS

Adaptations can be of three types—structural, physiological or behavioural.

- **Structural**—the physical features or structures of an organism.

Example: Mangroves have specialised aerial roots for oxygen uptake at low tides; they disperse buoyant seeds that germinate while still attached to the parent plant.

- **Physiological**—the way in which an organism functions.

Example: Flowering of the snow daisy occurs in summer; it can survive low winter temperatures.

- **Behavioural**—the way in which an organism acts or behaves.

Example: Some animals hibernate during cold periods.

The adaptations that organisms have affect their ability to survive in particular environments. In turn, their distribution is determined by the presence or absence of features that would make them well suited to a particular environment.

The environment of an organism must provide it with the things it needs—its requirements. Requirements include nutrients such as oxygen and water, as well as shelter. If a requirement is in short supply it can affect the survival and reproduction of an organism. This is called a limiting factor. For example, if the magnesium concentration in soil is too low to meet the needs of a plant (in this case the manufacture of chlorophyll), then the plant may not survive. Magnesium is a limiting factor, even if all of the plant's other requirements are met.

All organisms have a **tolerance range** within which they can survive. When conditions move beyond these limits the organism cannot survive. For example, a reptile with a temperature tolerance range of 0°–35°C will survive when this temperature range is met. However, the reptile will die if it is subjected to temperatures outside this range.

Organisms have tolerance limits for a range of factors.

EXAMPLES OF ADAPTATIONS

Aquatic plants

Aquatic plants (hydrophytes) are plants that live either partially or fully submerged in water. This presents problems in acquiring some nutrients and light and oxygen can be limiting factors. Hydrophytes have adaptations that assist them in overcoming such problems (Table 3.1).

Mangroves are plants found in the warm tidal waters of much of the Australian coastline. They must cope with a wide range of harsh environmental factors including excessive exposure to salt and inundation of their roots by water. Pneumatophores (aerial roots) are a structural adaptation that enhances oxygen uptake (Figure 3.3).

TABLE 3.1 Adaptations of aquatic plants (hydrophytes)

Adaptation	Example
flat leaves, few stomata, thin cuticle (submerged plants)	• marine sea grass • fish tank plants
emergent roots that protrude above the water	• cumbungi • reeds
roots under water, leaves above, have air-filled spaces to circulate air	• rice
floating leaves with stomata on upper surface of leaves	• water lily
free-floating plants feature air-filled spaces and thick cuticle	• duckweed

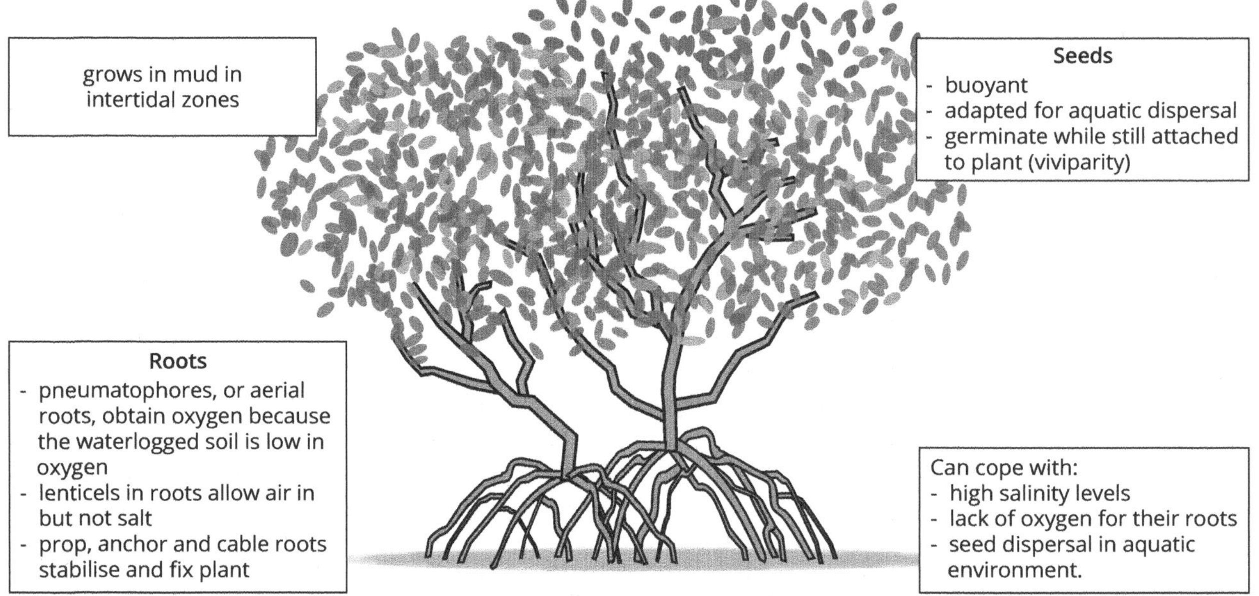

FIGURE 3.3 Specialised structures of mangroves

Plants in arid environments

Xerophytes are plants that are adapted to hot, dry environments (Figure 3.4). They have adaptations that reduce water loss and thus have an increased tolerance to dry conditions.

Adaptations in xerophytes include:

- a thick cuticle—reduces evaporation of water from the leaf surface
- hairs covering leaves—increase humidity around stomata thereby decreasing the water vapour gradient and reducing the diffusion of water vapour from the stomata to the external air
- sunken stomata—creates pockets of humid air around stomata, decreasing the water vapour gradient and reducing the diffusion of water vapour from the stomata to the external air
- fewer stomata—reduces sites from which to lose water
- fewer leaves—decreases surface-area-to-volume ratio, resulting in smaller surface across which to lose water by evaporation
- leaves facing away from the sun— reduces temperature of leaves and thereby reducing evaporation.

There are two types of xerophytes:

- succulents (fleshy, thick-stemmed), for example cacti
- sclerophylls (hard-leaved), for example eucalypts.

Succulents obtain and retain more water than non-succulent plants. They can have:

- downward-pointing spines to direct condensed water vapour downwards to drip on roots and deter herbivores
- no leaves (reducing surface area of plant to conserve water)
- specialised water storage cells in stem
- a thick cell epidermis to reduce evaporation
- a thin layer of cells for photosynthesis
- globular or rod-shaped leaves (decreasing surface-area-to-volume ratio) covered with white hairs or wax.

FIGURE 3.4 Xerophytes, such as cacti, are plants that are adapted to hot, dry conditions.

 ISBN 978 1 4886 1931 1

Plants in salty environments

Halophytes are plants that are adapted to high levels of salt in their environment. They are able to survive because they have mechanisms for removing excess salt. Halophytes may remove salt by:

- excretion through salt glands
- shedding leaves that have a high salt content.

Plant adaptations to fire

Some Australian plants are able to survive the extreme temperatures that accompany bushfire. Adaptations that allow this include:

- lignotuber—swelling at the base of the stem or trunk that allows regeneration even after the shoot has perished, for example mallee eucalypts
- epicormic bud—bud tissue that lies beneath the bark of some trees from which new plant tissue grows after fire.

Animals in arid environments

The desert is a harsh environment with very little rainfall and extreme temperatures. Desert animal life is not abundant. Animals that live in the desert have adaptations to cope with the lack of water, the extreme temperatures and the shortage of food (Figure 3.5 and Figure 3.6).

Some of the adaptations shown by land animals in arid environment are listed in Table 3.2.

FIGURE 3.5 Australian hopping mouse

FIGURE 3.6 Thorny devil

Theory of evolution by natural selection

Our modern understanding of change in populations (**evolution**) has been the subject of changing theories over the centuries.

- Jean Baptiste Lamarck (1744–1829) proposed a theory of evolution based on the idea that organisms 'acquired' characteristics in response to a need, for example giraffes acquired a long neck because they constantly stretched it in an attempt to reach the highest leaves in trees.
- Charles Darwin (1809–1882) based his theory of evolution by **natural selection** on his observations of organisms' adaptations to their environment. In accounting for such adaptations, he proposed that organisms most suited or adapted to their environments have the greatest chance of surviving and reproducing.
- Alfred Russell Wallace (1823–1913) independently developed the idea of evolution by natural selection.
- Gregor Mendel (1822–1884), an Austrian monk, undertook breeding experiments with pea plants and bees, meticulously recording his observations. The patterns of inheritance that he observed led him to make accurate predictions about the way in which characteristics are inherited from one generation to the next.

TABLE 3.2 Adaptations of animals to arid environments

	Adaptation		
	Structural	**Physiological**	**Behavioural**
Low water availability	• relatively round body shape reduces surface-area-to-volume ratio and thus reduces water loss, e.g. water-holding frog *Cyclorana*	• can tolerate high body temperatures • does not pant or sweat • does not drink—gets sufficient water from food and body metabolism • excretes uric acid (reptiles and birds) • has an accelerated life cycle during wet seasons • will hibernate • will produce concentrated urine (mammals)	• reduced activity • nocturnal activity • licks fur • spends time in shade
High temperatures	• small body size • covering of thick coat of insulating fur, exoskeleton or shell • grooves in skin to collect water	• excretes almost-dry faeces • has a rich network of capillaries in mouth and nose that increases heat loss • hibernates	• changes body posture and orientation to maximise heat loss • nocturnal activity—avoids the heat of day • lives in burrow where temperatures are lower

ISBN 978 1 4886 1931 1

TABLE 3.3 Change in population over time and generations

Variation exists within populations. Some individuals are more suited (adapted) to the environment; others are less well adapted.	
↓	↓
Individuals with favourable characteristics are said to be well adapted to their environment.	Individuals with less favourable characteristics are less well adapted.
↓	↓
Individuals have a survival advantage and so have an increased chance of reproducing and passing on their 'favourable' alleles.	Individuals have a decreased chance of surviving and reproducing compared to those with favourable characteristics.
↓	↓
Individuals leave a greater number of offspring that also have favourable alleles, i.e. offspring that are adapted to environment.	Individuals leave fewer offspring
↓	↓
There is an increase in the frequency of favourable alleles in the population.	There is a decrease in the frequency of the unfavourable alleles in the population.
↓	↓
There is an overall increase in the proportion of individuals with favourable characteristics in the population, and overall decrease in the proportion of individuals with less favourable characteristics.	
↓	↓
The population becomes increasingly adapted to its environment.	

Today we accept the theory of evolution by natural selection to account for changes in populations over time. Natural selection is a process in which individuals with the most favourable characteristics have a better chance of surviving and reproducing than individuals with less favourable characteristics. Individuals with characteristics that are well suited to their environment are biologically fit individuals and are 'selected for' (survive, reproduce and pass on their favourable alleles).

An adaptation is a feature displayed by an organism that makes it well suited to its particular environment. For example, wings are an adaptation for flight, and colour can provide camouflage—an adaptation to avoid predation.

Natural selection depends on variation in the **phenotypes** within a population. Populations show variation in the form of traits. A trait that has many forms is called polymorphic.

Variation in populations is generated as a result of mutation and sexual reproduction.

Sexual reproduction is the production of genetically unique gametes in meiosis (the process in which gametes/sex cells are produced) and the random union of those gametes in fertilisation, which further mixes the genetic material.

A **mutation** is a change in the DNA of an organism. Mutations introduce new alleles into populations, which may result in a change of phenotype. Mutations are the raw material for change in a species over time (evolution).

Table 3.3 summarises change in a population over time and generations.

Artificial selection occurs when humans intervene in the breeding of organisms such as plants and animals to deliberately select those individuals with the most desirable characteristics.

SPECIES CHANGE OVER TIME

The phenotype is the physical expression of an organism's genetic make-up or **genotype**. Both the genotypes and phenotypes present in populations change over time. This is usually most evident over many generations, especially in larger organisms with long life cycles. Change in the population of smaller organisms that have very short life cycles, such as insects, is often more easily observed. Environmental selection pressures act on the phenotypes of organisms.

Changes in the proportions of phenotypes present in populations occur as a consequence of changing **allele** frequencies (**gene** variants) over generations. Allele frequencies within populations change as a result of environmental pressures that act on phenotypes. Emigration and immigration also alter allele frequencies within populations. Emigration can reduce the frequency of specific alleles in a population. Immigration can introduce new alleles. Birth and mortality rates also change the variation within populations.

Genetic variation

- Gene: a section of DNA that carries the code for making proteins.
- Allele: a different form of a gene (gene variant).
- Genotype: the genetic make-up of an individual—the particular alleles present for a given gene or genetic marker.
- Phenotype: an individual's physical characteristics—the expression of the genetic make-up or genotype. The genotype, together with the environment, determine an individual's phenotype.
- Allele frequency: the proportion of a particular allele in a population.
- Gene pool: the genetic make-up of a population; includes the sum of all of the alternative alleles for different genes present in a population.

Selection pressures in the environment act on the phenotypes of individuals. As a result, the frequency of alleles in the population, and the gene pool, may change.

Population change and allele frequencies

Populations are subject to change as a result of various factors—gene flow, genetic drift, the founder effect and genetic bottlenecks.

Gene flow

The exchange of alleles between populations is known as **gene flow**. This occurs as a result of migrating individuals who reproduce in the new populations they join. Gene flow in plants occurs through the dispersal of pollen and seed. Gene flow can introduce new alleles into a population, changing its genetic composition.

Example: The movement of Fyo, Fya and Fyb alleles of the Duffy blood group in humans between populations of African and European origin. The original European settler population in the United States showed a high frequency of the Fya allele and the native African population a low frequency. Today, gene flow accounts for an overall increase of this allele in African–Americans.

Genetic drift

The change in allele frequencies in a population over generations as a result of chance alone is known as **genetic drift**. That is, no selection pressures appear to be acting on the phenotypes present. Small, isolated populations are more susceptible to genetic drift than larger populations or populations that maintain gene flow with neighbouring populations.

Example: A new mutation that is 'neutral', that is, does not put the individual at a survival disadvantage, may become fixed in a population, rapidly changing the genetic profile of the population.

Founder effect

Founder effect occurs when a small population of individuals is isolated and forms the basis of a new population. A founder group of individuals may have limited variation, and not be representative of the population from which it originates. After successive generations, the population is phenotypically different from its parent population.

Genetic bottleneck

Genetic bottlenecks occur when a large population is dramatically reduced in size, thereby reducing its genetic variation (also known as **genetic diversity**). The frequency of some alleles in the population may thus be reduced or eliminated altogether. Such events may reduce the chances of survival of a population or species, especially if favourable phenotypes are poorly represented in the remaining population.

Examples: Catastrophic events such as drought and volcanic activity can dramatically reduce population sizes. A meteorite strike is one theory that has been proposed to explain the bottleneck that occurred in dinosaur species approximately 65 million years ago, eventually leading to their extinction.

The helmeted honeyeater and Leadbeater's possum are examples of Australian species that have experienced genetic bottlenecks following the loss of habitat. Both these species are currently listed as Critically Endangered.

Microevolutionary change

Microevolutionary change refers to small-scale changes within species, such as the differences in allele frequencies of populations from one generation to the next. Changes in allele frequencies over many generations lead to cumulative changes in populations that can result in speciation; that is, new species diverging from ancestral ones.

Examples: The frequency of alleles conferring antibiotic resistance is increased from one generation of bacteria to the next; the frequency of alleles conferring resistance to myxomatosis is higher in subsequent generations of a rabbit population.

Macroevolutionary change

Macroevolutionary change refers to large-scale changes in the evolution of organisms that occur above the species level.

Example: Modern species of Australian marsupials that have diverged from ancestral marsupial species.

SPECIATION

A group of organisms is considered a distinct species if its members are similar and are able to interbreed in their natural environment to produce viable offspring.

Speciation is defined as the development of new species from pre-existing species.

Speciation occurs when populations become isolated from one another and no gene flow occurs between them. The different populations are subject to different environmental pressures. Over time and many generations, the populations diverge so that individuals are no longer able to interbreed, even if they do come together again.

Modes of speciation

Several models account for the evolution of new species from ancestral species. Each takes into account the factors acting on populations.

The most common form of speciation is **allopatric speciation**, which occurs when a population becomes divided by a geographical barrier. Populations are physically separated from one another so gene flow between the populations ceases, resulting in **reproductive isolation**.

Other forms of speciation include peripatric speciation, sympatric speciation and parapatric speciation.

- **Peripatric speciation** occurs when a small population becomes isolated at the perimeter or edge of the range of the parent population and gene flow is interrupted. When alleles present in low frequency in the parent population are present in the isolated group, they can become more common, quickly altering the allele frequency in the new population (founder effect). Eventually the two populations develop into distinct species.

- **Sympatric speciation** occurs when a population becomes divided so that gene flow is interrupted between the two groups within the same geographic area; usually these are populations inhabiting different microhabitats within the same larger geographic area. Eventually the population change results in separate species.
- **Parapatric speciation** is a rare form of speciation that occurs when gene flow is reduced or halted, not by any physical barrier, but due to factors that result in a population occupying a different **niche** that is adjacent to the one occupied by the parent population.

Speciation may be gradual or rapid.

Gradual speciation occurs when a population is divided into two approximately equal and representative groups that become isolated from one another. Selection pressures act on the phenotypes in the two populations according to the allopatric model above. Changes in the two populations eventually lead to the development of races and then distinct species. The process takes a great deal of time and occurs over many generations.

Rapid speciation may occur when a small, non-representative group becomes isolated from the original population. Advantageous mutations become fixed in the population and rapidly increase in frequency. The new group becomes different from the parent population relatively quickly, leading to speciation over a much briefer period of time.

Adaptive radiation is a form of speciation in which an ancestral population is divided into a number of populations that diverge to form different species. The 13 finch species of the Galapagos Islands are a cluster of related species that represent an example of adaptive radiation.

Figure 3.7 illustrates a model of some key steps in population change leading to speciation.

Isolating mechanisms between species

Similar, related species that share a particular habitat do not interbreed in their natural environment. A number of mechanisms appear to maintain reproductive isolation between species.

Prezygotic isolating mechanisms

Prezygotic isolating mechanisms are factors that prevent a **zygote** forming in the first place. For example, members of different species do not come into contact for breeding or when they do, other factors operate that prevent fertilisation even if mating were to occur.

- Geographic isolation—occurs due to the separation of populations by physical and geographic barriers such as oceans, deserts, mountain ranges and glaciers.
- Ecological isolation (or niche partitioning)—occurs where populations occupy different niches within the same habitat because they have different requirements. The populations may be in relatively close proximity, for example different species of eucalypt growing on a mountain slope but one at a higher altitude than the other.
- Temporal isolation—occurs when the timing of activities does not overlap, for example breeding cycles occur at different times of the year, or one species is nocturnal while another is diurnal.
- Behavioural isolation—the behaviours of members of different species mean that they do not recognise one another for the purposes of mating, or they do not come into contact for other reasons. For example, related species of frogs may have different mating calls that only members of the same species recognise.
- Structural or morphological isolation—structural differences in the genitalia of the two species in question may make interbreeding impossible.

Postygotic isolating mechanisms

Postzygotic isolating mechanisms are those that typically prevent a zygote of two different species from developing into a fertile adult. The offspring resulting from interbreeding between individuals from different species are called hybrids.

- Hybrid sterility—if mating does occur, the offspring are infertile and no further offspring can be produced. Because fertilisation actually occurs and the hybrid individual develops into a mature organism, hybrid sterility is a postzygotic isolating mechanism.
- Chromosomal abnormalities:
  - **polyploidy** is the failure of whole sets of paired chromosomes to separate during meiosis and can give rise to new cells with multiple sets of chromosomes. This phenomenon is relatively common in some plants. The polyploid individuals represent a different species in terms of their reproductive compatibility with the parent species from which they arose. Hence, polyploidy is a form of instant speciation.
  - **aneuploidy** refers to the abnormal number of chromosomes in a cell, that is, not a multiple of the haploid number for the species in question. This condition occurs as a result of chromosome pairs failing to separate during meiosis (nondisjunction) and leads to some cells containing too many chromosomes and others too few. Aneuploidy is rare and problematic in animals, but more common in plants and yeasts. It is responsible for introducing multiple new phenotypes into a population. Aneuploidy in yeast cells is thought to be linked to rapid adaptive evolution.

PATTERNS OF EVOLUTION

The interactions between the environment and the phenotypes within a population are numerous, complex and change over time. Because of this complexity, the species we observe today have experienced many different selection pressures and patterns of evolution over time. Scientists consider all of the information available about organisms, both extant (living) and extinct (see fossil information on pages 109–110), as well as evidence from the environment (selection pressures) to construct an understanding of how organisms have changed over large expanses of geological time.

 ISBN 978 1 4886 1931 1

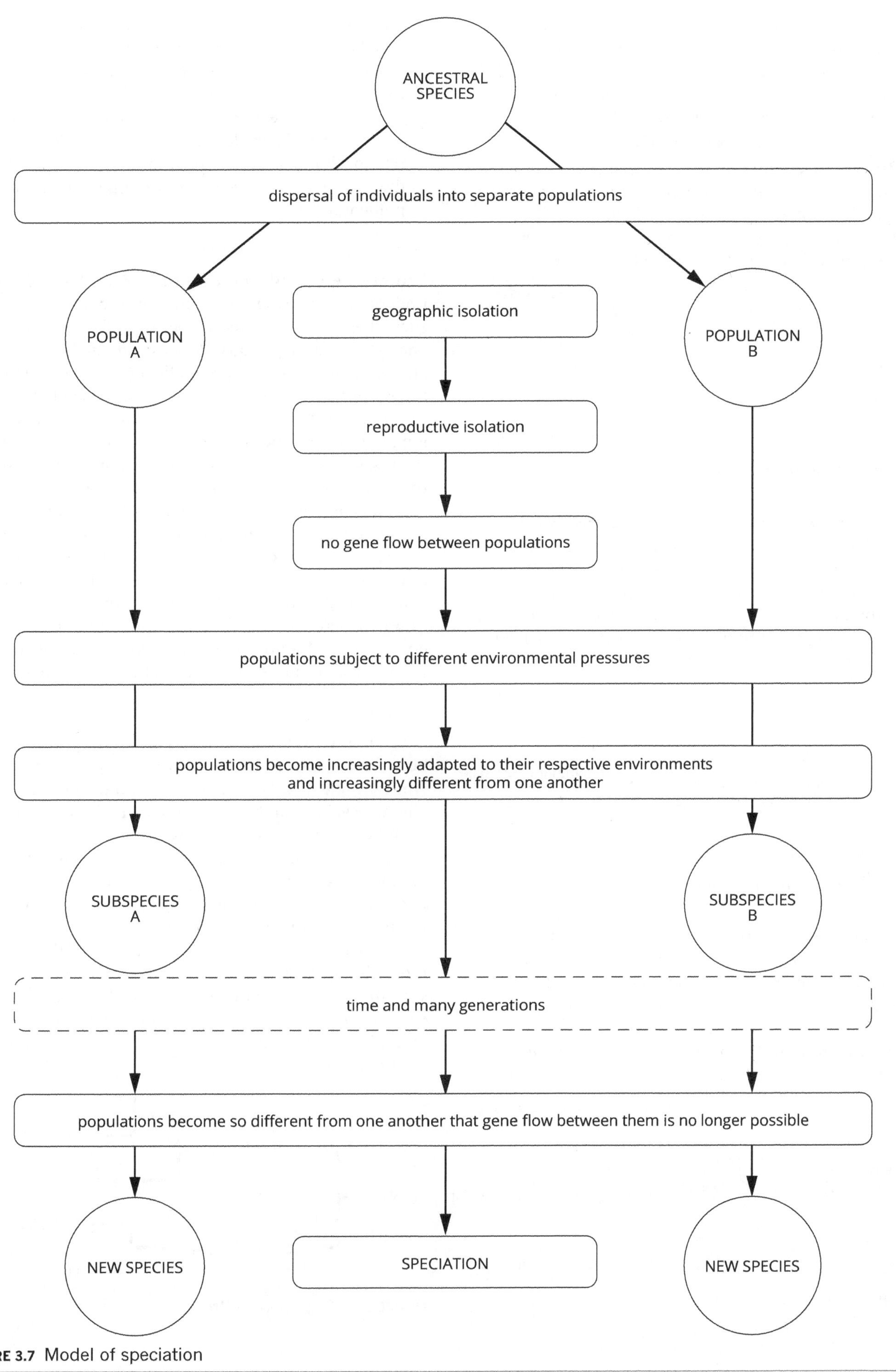

FIGURE 3.7 Model of speciation

Comparing organisms to similar and different species is important in understanding the evolutionary relationships between organisms. See pages 111–112 for other evidence used to establish evolutionary relationships, for example biochemical evidence.

There are four main patterns of evolution:

- **Divergent evolution**—species that share a common ancestral species have diverged over time, evolving into distinctly different species (Figure 3.8). For example, the body shape and flippers of dolphins and seals are evidence of their common ancestry.

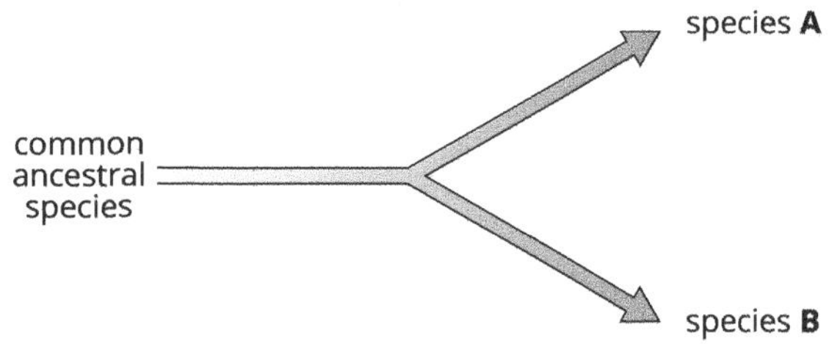

FIGURE 3.8 Divergent evolution

- **Convergent evolution**—unrelated species display similar features as a result of similar environmental pressures but not shared ancestry (Figure 3.9). For example, the similar body shape of dolphins (a mammal) and sharks (a fish) is evidence that they have experienced similar environments and selection pressures but are not closely related species.

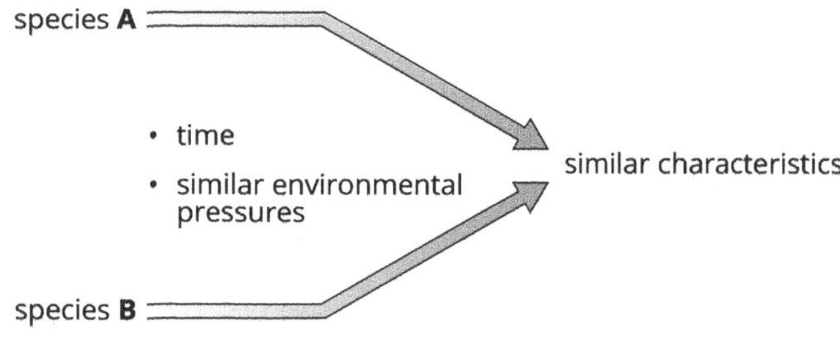

FIGURE 3.9 Convergent evolution

- **Parallel evolution**—related species that share a common evolutionary ancestor are subsequently subject to similar environmental pressures that lead to similar features (Figure 3.10). For example, the similarities in wing patterns and colouration of different butterfly species that live in similar environments is an example of parallel evolution.

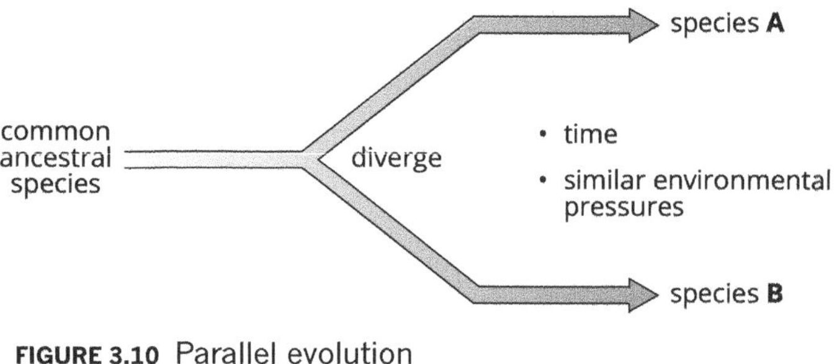

FIGURE 3.10 Parallel evolution

- **Coevolution**—refers to species that interact so closely that they exert selection pressures on each other. Essentially, such species rely on one another for their survival. They evolve together so that their responses to selection pressures are reciprocal to one another. Examples of coevolution include the relationship between a predator and its prey and the relationship between a flowering plant and its pollinator.

Punctuated equilibrium

Punctuated equilibrium is the theory that species undergo rapid change from one form to another, with such changes punctuated by periods of stability or little change over long periods. Punctuated equilibrium is an alternative theory to gradualism (Darwin's original theory of natural selection), which proposed that evolution always occurs gradually and uniformly over long periods of time. The many transitional forms in the fossil record support the theory of gradualism but there are also many instances where species seem to suddenly appear with no evidence of transitional forms. Punctuated equilibrium explains why we cannot always see changes in species even though we know that they have occurred—there are long periods of time where no change occurs (the equilibrium) and then short periods of time where changes do occur (punctuating the equilibrium). Rapid evolutionary change usually occurs in response to sudden changes in the environment.

Phylogenetic trees

A **phylogeny** is the pattern of evolutionary relationships between organisms. Evidence from various sources (for example, fossils, morphology and molecular characteristics) can be used to build a picture of the evolutionary relationships between different species. Such relationships can be represented in diagrams called phylogenetic trees (Figure 3.11 and Figure 3.12). The position of an organism in a phylogenetic tree represents its evolutionary relationship to other organisms in the tree. The closer organisms are positioned in the tree, the more closely related they are. In Figure 3.12, it can be seen that birds and reptiles are more closely related to each other than to any other group in the phylogenetic tree. This indicates that birds and reptiles share a recent common ancestor.

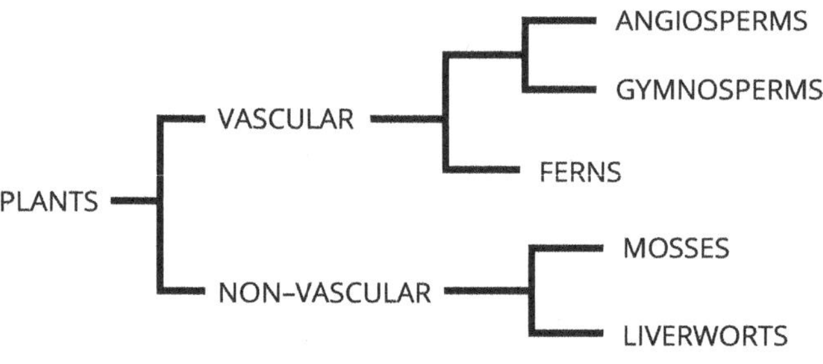

FIGURE 3.11 Phylogenetic tree of plants

 ISBN 978 1 4886 1931 1

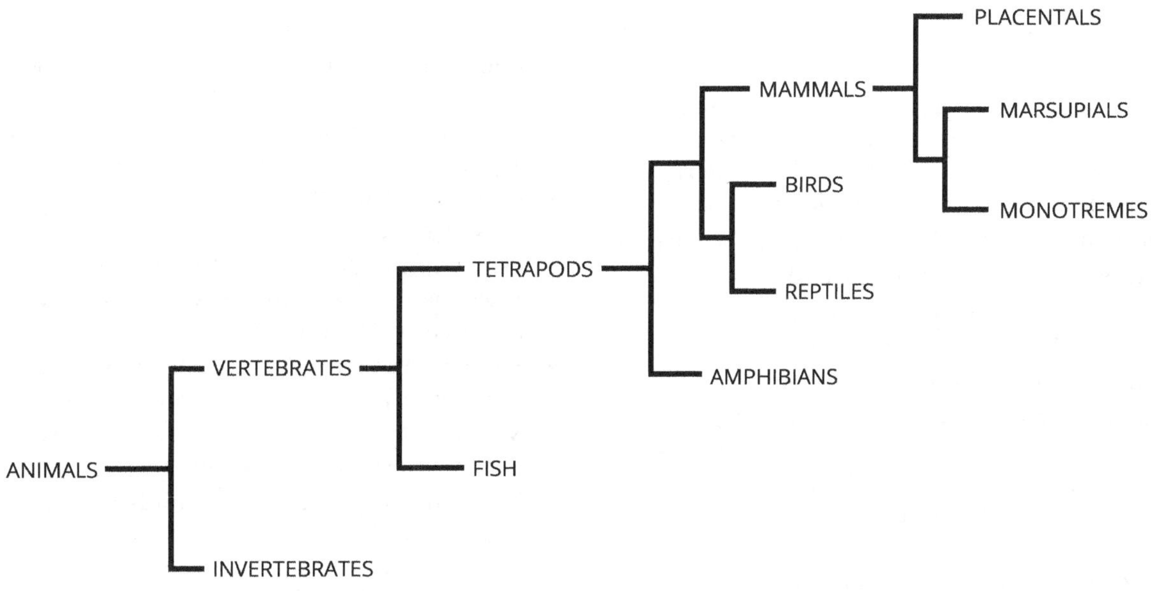

FIGURE 3.12 Phylogenetic tree of animals

Evolution—the evidence

Evolution is the process of change in living organisms over time. The modern theory of evolution states that all living organisms share a common origin that dates back to around 3.8 billion years ago. Evidence supporting the theory that life on Earth formed through a process of evolution is present in a number of forms: the fossil record, biogeography, comparative anatomy, comparative embryology, and biochemical evidence such as DNA sequences.

FOSSILS AND PALAEONTOLOGY

Fossils provide strong evidence supporting the theory of evolution. A fossil is the remains or traces of an organism that lived before the present time.

Fossils exist in different forms:

- Preserved remains—all or part of an organism remains intact (e.g. mineralised bones and shell insects trapped in tree resin, mammoths preserved in ice sheets).
- Impression fossils—dead remains of organisms become encased in rock; the remains dissolve but an imprint of the organism is left in the rock (e.g. leaf impressions in rock).
- Casts—dead remains encased in rock dissolve away and the space left becomes filled with foreign material, for example silt, that forms a 3D replica of the organism (e.g. Gogo fish in the Kimberley, Western Australia).
- Traces—indirect evidence that an organism existed (e.g. footprints, nests and scats).

Fossilisation of organisms is a relatively rare occurrence, but when it does occur it is most effective when certain conditions are met. These include a lack of oxygen, which reduces the rate of decay by bacteria, and a lack of scavengers. For these reasons, aquatic organisms are well represented in the fossil record.

Example: fish dies and sinks to sea floor

↓

water currents cover dead remains with sand before scavengers reach it

↓

depth of sea floor and burial by sand reduce oxygen availability for decomposition by bacteria

↓

weight of water and sediments apply pressure to dead remains, replacing hard tissue such as bones with minerals.

The fossil record indicates that:

- life forms have existed on Earth for over 600 million years
- there has been an increase in the complexity of life forms over time
- there has been an increase in **biodiversity** (the variety of life forms) over time
- extinction of species has also occurred and continues to occur. **Mass extinctions** are extinction events characterised by the disappearance of a large number of species in a relatively short timeframe. Five major extinction events are recognised in the fossil record. We are currently experiencing the sixth major extinction event
- life forms have existed that are indicative of significant evolutionary changes between species, for example fossils with features that suggest a partially aquatic and partially terrestrial lifestyle. Such fossils are called transitional fossils because they represent life forms that are intermediate or transitional between distinctly aquatic and distinctly terrestrial organisms.

Dating techniques and the geological time scale

The fossil record shows the emergence of different kinds of living organisms set against geological time. Geological time is a measure of the history of Earth in terms of its geology, and is divided into five eras:

- Archaean: oldest rock, 4600–2500 million years ago
- Proterozoic: 2500–541 million years ago
- Palaeozoic: 541–252 million years ago
- Mesozoic: 252–66 million years ago
- Cenozoic: 66 million years–present time

The earliest known fossils have been dated as Precambrian and include stromatolites (prokaryotes) and multicellular Ediacaran fauna. Trilobites are examples of early Palaeozoic life forms.

Various dating techniques are available to estimate the relative and actual age of rock and fossils.

- **Stratigraphy**—comparison of rock strata and the fossils they contain. The oldest rock strata were laid down first; upper layers are more recent. The relative age of a fossil can be determined by comparing the position of the rock stratum in which it was found with rock strata above and below.
- **Index fossils**—useful in dating unidentified fossils. Features of index fossils include: they have been identified; they are common in rock strata of a particular age; the age of the rock in at least one location is known. Fossils discovered in a new location that are either similar to the index fossil, or have been found with fossils of the same kind as the index fossil, suggest that the rock stratum is the same age. Index fossils are also called indicator fossils.
- Absolute dating techniques—used to determine with accuracy the actual age of rock or fossils. Absolute dating techniques measure the decay rates (half-life) of radioactive isotopes to assess age. This is called **radiometric dating**. The half-life of an isotope is the time taken for half of the atoms in the radioactive sample to decay (undergo chemical change).

TABLE 3.4 Radiometric dating techniques

Radiometric technique	Isotopes	Examples
Radiocarbon dating	measures ratio of C^{14} to C^{12}	limited to dating fossils less than 50000 years old
Potassium–argon dating	K^{40} : Ar^{40}	4600 million years old
Uranium–lead dating	U^{238} : Pb^{207}	4600 million years old

- Thermoluminescence—used to date human artefacts, for example cooking utensils. It measures light emissions from minerals when heated. The intensity of light emitted from such objects provides a measure of the amount of time that has passed since the object was heated.

BIOGEOGRAPHY

Biogeography is the study of how organisms are distributed geographically. A logical explanation to account for the presence of similar species on different continents is that the continents were once connected as a single, large landmass.

Observations:

- Galaxiid fish, that live in freshwater, are present in Australia, New Zealand, New Caledonia, South America and South Africa.
- Different but related species of baobab trees are found in Africa, Madagascar and Western Australia.
- Various species of *Waratah* are native to Australia, South America and New Guinea.
- Different species of flightless birds (ratites) are native to Australia, New Guinea, New Zealand, South America and Africa.

The theory of continental drift explains this phenomenon.

Explanation:

- The continents are parts of crustal plates that make up Earth's rocky surface and these plates move in relation to one another.
- More than 180 million years ago, the landmasses of Australia, New Zealand, Antarctica, India, Africa, Madagascar and South America formed a single supercontinent called Gondwana.
- Movement of crustal plates, also called tectonic plates, has resulted in a break-up of this supercontinent.
- Over the period 132-96 million years ago Australia separated from Gondwana.
- Australia completed its separation from Antarctica approximately 30 million years ago.
- The shapes of the continents also show contours that reflect how they may have once fitted together.

Conclusion: ancestral species of different organisms that already existed became isolated on their respective continents and continued to evolve to their present forms.

COMPARATIVE ANATOMY

Comparative anatomy examines the structural similarities and differences between different groups of organisms. It poses the question: why do different kinds of organisms possess similar basic anatomy?

Observation: the limbs of different groups of tetrapods display similar or homologous skeletal structures (Figure 3.13).

Conclusion: the different groups of organisms are related and share a common ancestor, from which these different groups diverged.

ISBN 978 1 4886 1931 1

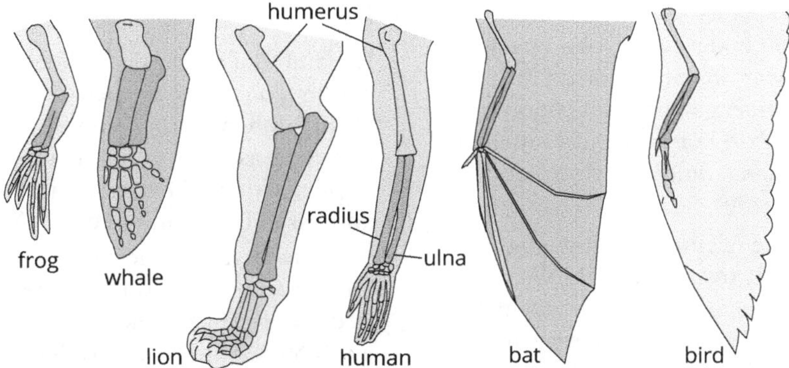

FIGURE 3.13 The bones in the forelimbs of tetrapods are homologous.

- **Homologous** features—fundamentally similar in morphology, for example the wings of bats, hands of humans and flippers of seals show similar basic bone structure and arrangement but different proportions related to different functions. Homologous structures are indicative of common evolutionary ancestry.
  - **Vestigial** structures—appear in reduced form in one species compared to a more developed form in another, for example the human appendix is the same anatomical structure as the end pocket in the caecum of herbivorous mammals. Its presence in humans suggests a function in early human evolution and is evidence that humans share a common ancestry with herbivorous mammals.
- **Analogous** features—perform the same function but are fundamentally different in structure, for example the wings of butterflies and birds. Analogous structures suggest that the organisms are not closely related. Such features are the result of convergent evolution.

COMPARATIVE EMBRYOLOGY

Comparative embryology examines and compares the patterns of embryological development of different species (Figure 3.14).

Observation: embryological development in different groups of vertebrates follows the same pattern.

Conclusion: the same development pattern suggests the different groups share a common evolutionary ancestor.

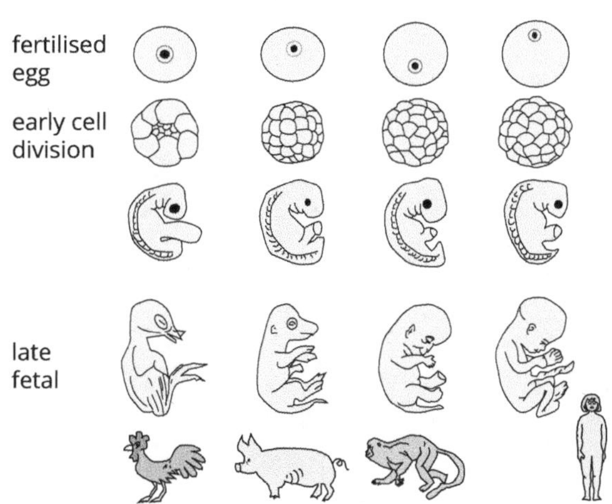

FIGURE 3.14 Embryological development in vertebrates

BIOCHEMICAL EVIDENCE

Biochemical evidence includes various aspects of the biochemistry of living things, such as DNA, RNA and proteins, which provide evidence of the evolutionary relationships between species.

- Homologies between similar chromosomes in different related species can be useful in establishing the relatedness of the species.
- Amino acid sequences in polypeptide chains of some proteins are identical between some species, for example cytochrome c and haemoglobin in humans and chimpanzees.
- **DNA sequencing** is used to compare the degree of similarity in the DNA sequences or genomes of different species (comparative genomics). Because DNA accumulates mutations, over time the DNA of isolated populations becomes different—what was once similar DNA gradually diverges. The more mutations observed in the DNA sequences of two species, the more time has passed since the two species diverged from their common ancestor. Various sources are available for DNA sequencing—nuclear DNA, mitochondrial DNA, chloroplast and ancient DNA.

TABLE 3.5 Comparison of DNA of different species using DNA sequencing

DNA sequencing: % similarity between human genome and	
chimpanzee	97.6
gibbon	94.7
capuchin monkey	84.2

The high degree of similarity revealed by DNA hybridisation suggests a close evolutionary relationship between species.

- Ancient DNA refers to DNA extracted from fossils that remains useful in comparing the DNA of related modern species.
- Mitochondrial DNA undergoes change (mutation) rapidly compared to nuclear DNA. Comparing mtDNA of related species is a useful tool in estimating the timeframe for the divergence of species that have separated in evolutionary terms over a relatively short time (about 20 million years). Estimation of time since the divergence of mammal groups is a relevant application.

- Chloroplast DNA in plants and algae is a short, circular molecule that undergoes evolutionary change relatively slowly, making it useful in estimating when divergence occurred between groups over a large period of geological time, for example flowering plants (angiospermae) and conifers (gymnospermae).

- **Molecular clock**—rate of mutational change observed in the different sources of DNA molecules of species over time. It can be used to measure the degree of genetic similarity and difference between species and to estimate the time since related species diverged. This provides a picture of the degree of evolutionary relationships between species.

In summary, biochemical similarities between different species, such as shared genes or DNA sequences, can provide strong evidence that species share recent common ancestors. This helps scientists understand the evolutionary history of groups of organisms.

RECENT EVOLUTIONARY CHANGE

Evolution of species is a phenomenon that has continued as long as species have existed. Changes in allele frequencies in response to selection pressures in populations, or in combination with factors such as immigration and emigration, represent some examples addressed in this module.

A modern and relevant example that has significant consequences for humans is represented by mutations in disease-causing bacteria that confer a resistance to antibiotics. Since the introduction of antibiotics to treat bacterial infections in the 1940s, bacteria resistant to antibiotics have continued to arise. Antibiotics present bacteria with a strong selection pressure—those bacteria that are resistant to antibiotics will survive while those that are not resistant will die. The antibiotic resistant bacteria continue to reproduce, increasing the number of bacteria that are resistant. The rapid rate of reproduction and short generation times of bacteria allow antibiotic-resistant bacteria to increase exponentially. The rapid evolution of antibiotic-resistant bacteria has serious implications for human health and disease management. It has challenged scientists to approach the development of antibiotics in new and innovative ways.

ISBN 978 1 4886 1931 1

WORKSHEET 3.1

Knowledge review—the language of biological diversity

1 The lists in the following table are divided into abiotic and biotic factors. Look for similarities in the items in each list.

Discriminate between the two terms by writing a definition for each:

a abiotic factor

b biotic factor

Abiotic		Biotic	
air water rock rain snow	lava earthquake lightning cloud sand	spider grasshopper earthworm mushroom lichen	moss fern gum tree kookaburra koala

2 Living organisms in every ecosystem on Earth are constantly facing environmental pressures that challenge survival of the individual and sometimes the species. Such environmental pressures are referred to as selection pressures.

Individuals with features (adaptations) that help them survive these pressures are 'selected for'—they have an increased chance of survival, reproduction and passing on their beneficial traits to their offspring. Individuals are vulnerable if they do not have features that help them survive these pressures. They are described as being 'selected against'. Selection pressures in the environment of an organism interact with the features of organisms (phenotypes) and determine the populations that survive or become extinct.

Identify the selection pressure in each example described below.

	Scenario	Selection pressure	Effect on individuals	Possible consequences for population
a	albino pygmy possum living in the Australian bush			
b	*Eucalyptus* trees infected by cinnamon fungus (*Phytophthora cinnamomi*), which attacks the roots and kills the tree			
c	trophy hunting of black rhino (Critically Endangered)			
d	human populations living in areas infested with the *Anopheles* mosquito, which carries the deadly malaria protozoan			

3 Using references to refresh your understanding, define each of the following terms:

a adaptation

b biodiversity

ISBN 978 1 4886 1931 1

WORKSHEET 3.2

The cane, the beetle, the toad and the quoll—a tale of selection pressures and population change

Sugar cane (*Saccharum officinarum*) is native to Asia and Melanesia. It was brought to Australia with the First Fleet in 1788, but was not a successful commercial crop plant until 1862. In 1935 the native South and Central American cane toad (*Rhinella marina*) was introduced as a biological control mechanism to combat the problematic sugar cane beetle (*Dermolepida albohirtum*).

Facts about the cane toad:

- It did not successfully reduce cane beetle numbers.
- It preys on insects, snails and small vertebrates such as birds and marsupial mice.
- It has no natural predators that control its population in Australia.
- It is well suited to tropical Australian conditions.
- It is a prolific breeder.
- It secretes a deadly toxin from glands on its shoulders.

Other facts:

- The goanna (*genus Varanus*) and the Critically Endangered northern quoll (*Dasyurus halluctus*) are some Australian native carnivores that prey on the cane toad.

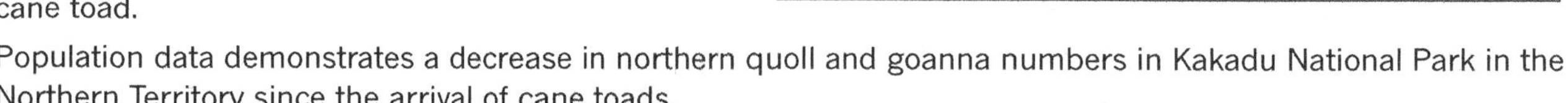

- Population data demonstrates a decrease in northern quoll and goanna numbers in Kakadu National Park in the Northern Territory since the arrival of cane toads.
- The cane toad disrupts the nests and eats the eggs of the the rainbow bee-eater (*Merops ornatus*), a native bird.

1 Use coloured pencils and the legend below to shade the distribution map of the cane toad in Australia.

1935–1974: yellow

1975–1980: green

1981–1986: red

1987–2001: orange

2002–2004: dark blue

Predicted distribution: light blue

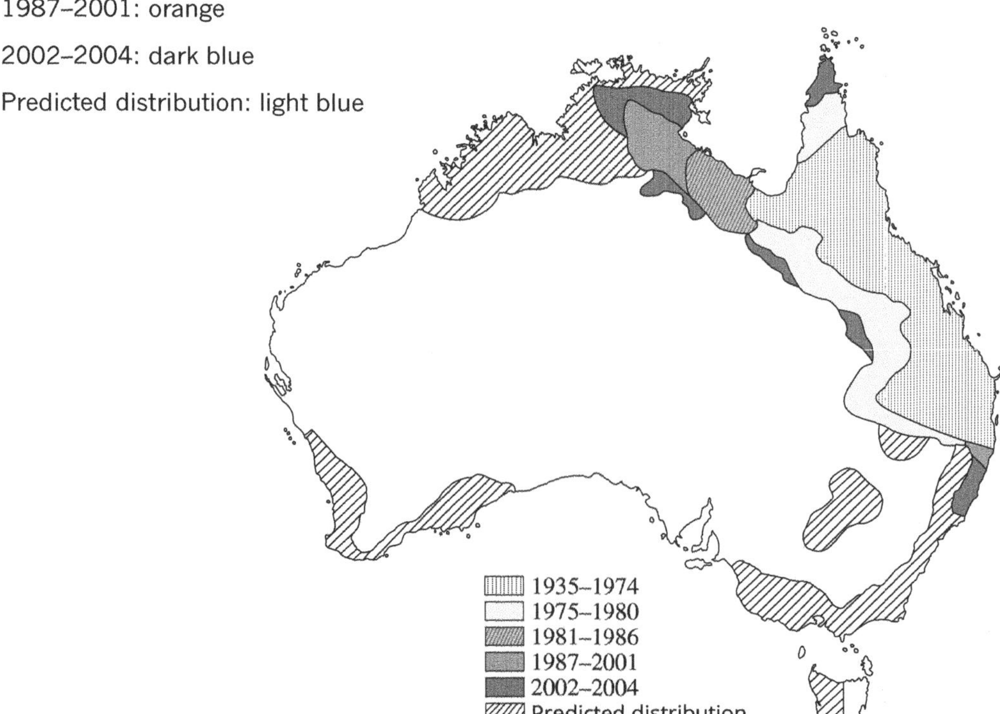

 ISBN 978 1 4886 1931 1

2 Give reasons to account for the increasing range of the cane toad in Australia.

3 Summarise the selection pressures and their impacts on each of the organisms listed in the following table.

Organism	Selection pressures	Impacts of selection pressures
cane beetle		
cane toad		
northern quoll		
rainbow bee-eater		

4 Explain what is meant by the phrase 'biological control'.

5 Predict a possible consequence for the northern quoll as the cane toad population continues to increase in Australia.

6 Summarise how environmental pressures have contributed to population change in Australian native species since the introduction of the cane toad.

RATING MY LEARNING	My understanding improved	Not confident ◄——► Very confident ○ ○ ○ ○ ○	I answered questions without help	Not confident ◄——► Very confident ○ ○ ○ ○ ○	I corrected my errors without help	Not confident ◄——► Very confident ○ ○ ○ ○ ○

ISBN 978 1 4886 1931 1

WORKSHEET 3.3

Amazing adaptations—features for survival

Organisms face many challenges in their environments. Features that help them to survive such challenges are called adaptations. These can be structural (for example, an anatomical feature), behavioural (related to the way an organism acts), or physiological (related to processes in the body). Consider the adaptations of the animals shown below that help them to survive in their particular environments. In each case:

- outline the challenge faced by the organism
- describe how the feature offers survival value
- identify the adaptation as structural, behavioural or physiological.

Goanna basking in sun

Challenge: ______________________________

Adaptation: ______________________________

____

____

____

structural behavioural physiological

Sugar glider with outstretched membranes

Challenge: ______________________________

Adaptation: ______________________________

____

____

____

structural behavioural physiological

Pelicans—salt glands at base of beak

Challenge: ______________________________

Adaptation: ______________________________

____

____

____

structural behavioural physiological

RATING MY LEARNING	My understanding improved	Not confident ◄──► Very confident ○ ○ ○ ○ ○	I answered questions without help	Not confident ◄──► Very confident ○ ○ ○ ○ ○	I corrected my errors without help	Not confident ◄──► Very confident ○ ○ ○ ○ ○

 ISBN 978 1 4886 1931 1

WORKSHEET 3.4

Elements of evolution—change in populations

There is a wide range of factors that contribute to evolutionary change in species.

1 Use a ruler and pencil to match each of the statements on the left with its corresponding example on the right.

Statement		Example
Variation exists within populations.		Two pink-flowering snapdragon plants produce some red-flowering and some white-flowering offspring.
Members of species show **adaptations** to their environments.		A dairy farmer deliberately retains cows for breeding that produce large volumes of milk.
Sexual reproduction results in offspring that are different from their parents and from one another.		Insecticide-resistant blowflies increase in a population exposed to the insecticide, while the frequency of sensitive individuals declines.
Natural selection results in an increase in the frequency of **biologically fit** individuals, i.e. those with features that are suited to the environment, compared to individuals that do not have these features.		Hair colour in humans is determined by the combinations of alleles each person possesses. This accounts for the range of hair colours, including black, brown, red, blonde and all colours in between.
Human intervention in the form of **artificial selection** can alter the genotype and phenotype of populations of organisms.		Migration from one population into another introduces new genetic variation into the second population.
The **gene pool** of a population is the sum total of all of the alleles present in the population.		Feather colouration in the tawny frogmouth makes it well camouflaged in its natural environment.
Dispersal of seeds and pollen by water, wind and animals leads to **gene flow** between populations of plants.		A population of rabbits has some individuals that are resistant to calicivirus and some that are sensitive.

2 Define each of the terms that appear in bold type.

RATING MY LEARNING	My understanding improved	Not confident ◄—► Very confident ○ ○ ○ ○ ○	I answered questions without help	Not confident ◄—► Very confident ○ ○ ○ ○ ○	I corrected my errors without help	Not confident ◄—► Very confident ○ ○ ○ ○ ○

WORKSHEET 3.5

Pathways of change—patterns in evolution

1 Evaluate the accuracy of each of the statements below. Circle either TRUE or FALSE for each statement.

Statement	
Evolution involves change in populations of organisms over time and generations.	TRUE/FALSE
Divergent evolution occurs when related species diverge from a common ancestor and then independently evolve similar features.	TRUE/FALSE
Convergent evolution occurs when unrelated species display similar adaptations. The presence of similar adaptations is attributed to different groups being subjected to similar environmental pressures.	TRUE/FALSE
The theory of punctuated equilibrium proposes that species undergo gradual change that is punctuated by short periods of stability.	TRUE/FALSE
Coevolution occurs when different species sharing the same environment, and interacting with one another, display similar adaptations as a consequence of similar environmental pressures.	TRUE/FALSE
The rapid divergence of one or more species from a common ancestral group is referred to as adaptive radiation. The 13 different finch species on the Galapagos Islands are an example of adaptive radiation.	TRUE/FALSE
Populations of the same species living in different geographic locations and subject to different environmental pressures are likely to show variation in the frequency of the same alleles.	TRUE/FALSE
When populations of the same species are geographically isolated over long periods of time and subject to different environmental pressures, the populations become increasingly adapted to their respective environments and increasingly different from each other. If members of the different groups become so different that reproduction is no longer possible, allopatric speciation is said to have occurred.	TRUE/FALSE
When paired chromosomes fail to separate during meiosis, offspring may have an extra set of chromosomes. This is a form of speciation called polyploidy. Polyploidy is common in some plant groups.	TRUE/FALSE
Mechanisms that maintain reproductive isolation between related species include behavioural mechanisms such as differences in mating calls, and structural differences in genitalia that prevent reproduction.	TRUE/FALSE
Genetic drift and the founder effect are processes that account for stability in some small isolated populations of organisms.	TRUE/FALSE

2 Examine each FALSE statement. Rewrite each as a TRUE statement.

RATING MY LEARNING							
My understanding improved	Not confident ◄──► Very confident	○	○	○	○	○
I answered questions without help	Not confident ◄──► Very confident	○	○	○	○	○
I corrected my errors without help	Not confident ◄──► Very confident	○	○	○	○	○

ISBN 978 1 4886 1931 1

WORKSHEET 3.6

Span of species—change on a grand scale

Earth is over 4500 million years old. There is evidence to indicate that life forms existed as early as 3500 million years ago. Since then the landscape of our planet has changed, and so too have the life forms that call it home. Australia in particular has been the cradle for the development of plants and animals that are unique in the world.

1 Use the geological timescale on the following page, together with other resources (for example, museum websites), to complete the table summarising key events in the evolution of life on Earth set against the approximate time in millions of years since these events are thought to have occurred.

<table>
<tr><th>Geological time period</th><th>Millions of years ago</th><th>Plants</th><th>Animals</th></tr>
<tr><td>Quaternary Period</td><td>2.58—present</td><td></td><td></td></tr>
<tr><td>Cretaceous Period</td><td>145–66</td><td></td><td></td></tr>
<tr><td>Jurassic Period</td><td>201–145</td><td></td><td></td></tr>
<tr><td rowspan="2">Precambrian Era</td><td>541</td><td>• algae diversifies</td><td>• animals with bodies protected by shells and first fish appear</td></tr>
<tr><td>4000</td><td colspan="2"></td></tr>
</table>

2 Find out what changes occurred in Australia over the last 110 million years. Briefly summarise these changes, including its global position in relation to other landmasses, climate, as well as flora and fauna. Draw diagrams to support your answer.

3 Referring to the figure below and your knowledge and understanding of evolutionary change, outline the pattern of change that has occurred in the abundance, complexity and diversity of species over geological time.

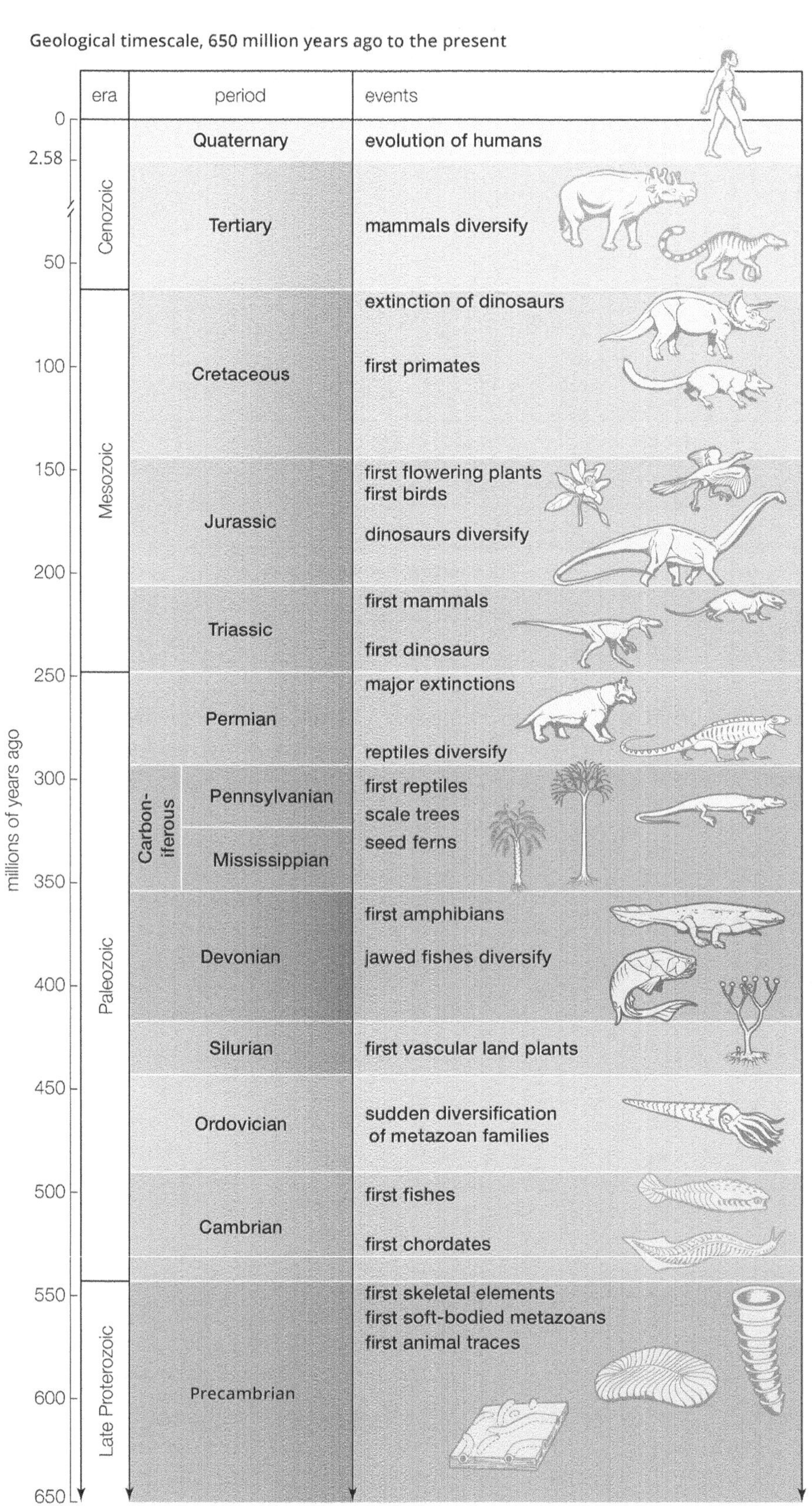

RATING MY LEARNING	My understanding improved	Not confident ◄—► Very confident ○ ○ ○ ○ ○	I answered questions without help	Not confident ◄—► Very confident ○ ○ ○ ○ ○	I corrected my errors without help	Not confident ◄—► Very confident ○ ○ ○ ○ ○

 ISBN 978 1 4886 1931 1

WORKSHEET 3.7

Homologous hands—comparative anatomy

Mammals, birds, reptiles and amphibians are very different groups of vertebrates. Even within each group there is an enormous diversity of forms, yet similarities are also evident.

1 The diagrams below represent the forelimbs of seven different vertebrates belonging to these groups.

- Compare the forearm anatomy of the different organisms to help you identify and label the bones that have not been labelled.
- Use a different colour pencil to colour each bone (for example, colour the humerus bone in each animal red).

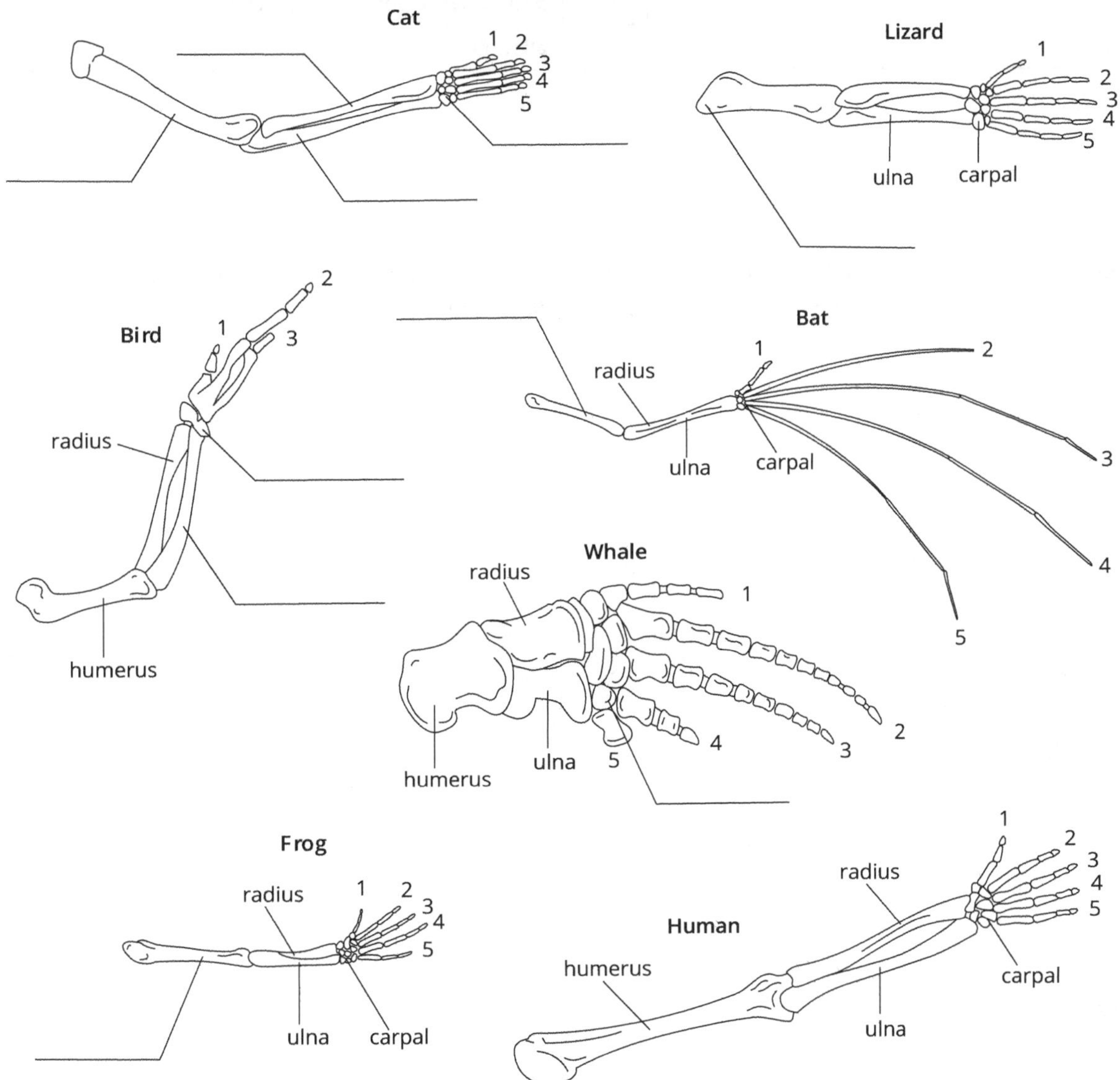

2 The forelimbs of these tetrapod vertebrates are called homologous structures. Explain what is meant by the term 'homologous'.

3 What do homologous structures suggest about the evolution of these different organisms?

RATING MY LEARNING	My understanding improved	Not confident ◄——► Very confident ○ ○ ○ ○ ○	I answered questions without help	Not confident ◄——► Very confident ○ ○ ○ ○ ○	I corrected my errors without help	Not confident ◄——► Very confident ○ ○ ○ ○ ○

WORKSHEET 3.8

Literacy review—evolution essentials

The terms we use in science often provide clues about their meaning. The parts of words work together to build richer meanings. The following table illustrates some of these words and parts of words relevant to Module 3: Biological diversity.

evolve: to change gradually, unfold	**disjoin:** disconnect, separate
micro: small, tiny	**ploidy:** refers to the number of sets of chromosomes in a cell
macro: large	**patria:** native home/homeland
poly: many	**allo:** other

1 Define 'species'.

2 Use your understanding of the words and parts of words to write a logical explanation of what each of the following terms mean so that a student just starting biology will understand them:

a evolution

b speciation

c allopatric speciation

d nondisjunction

e polyploidy

3 Use the terms below to briefly define microevolutionary change and macroevolutionary change. Describe an example of each from your study of this module.

population   allele frequency   generation

Terms	Definition	Example
microevolutionary change		
macroevolutionary change		

RATING MY LEARNING	My understanding improved	Not confident ◄——► Very confident ○ ○ ○ ○ ○	I answered questions without help	Not confident ◄——► Very confident ○ ○ ○ ○ ○	I corrected my errors without help	Not confident ◄——► Very confident ○ ○ ○ ○ ○

 ISBN 978 1 4886 1931 1

WORKSHEET 3.9

Thinking about my learning

On completion of Module 3: Biological diversity, you should be able to describe, explain and apply the relevant scientific ideas. You should be able to work with data, to interpret, analyse and evaluate it.

1 The following table lists the key knowledge covered in this module. Read each and reflect on how well you understand each concept. Rate your learning by shading the circle that corresponds to your level of understanding for each concept. It may be helpful to use colour as a visual representation. For example:

- green—very confident
- orange—in the middle
- red—starting to develop.

Concept focus	**Rate my learning** Starting to develop ◄——► Very confident				
Selection pressures in ecosystems that drive population change	○	○	○	○	○
Biotic and abiotic factors	○	○	○	○	○
Adaptations—structural, physiological, behavioural	○	○	○	○	○
Diversity of life over geological time	○	○	○	○	○
The theory of evolution by natural selection—Darwin and Wallace	○	○	○	○	○
Patterns of evolution—convergent evolution, divergent evolution, punctuated equilibrium	○	○	○	○	○
Evidence for evolution—fossils, biogeography, comparative anatomy, comparative embryology, biochemical evidence	○	○	○	○	○
Fossil dating techniques	○	○	○	○	○
Modern day examples of evolutionary change, e.g. antibiotic resistance of bacteria	○	○	○	○	○

2 Consider points you have shaded from starting to develop to middle-level understanding. List specific ideas you can identify that were challenging.

3 Write down two different strategies that you will apply to help further your understanding of these ideas.

PRACTICAL ACTIVITY 3.1

A numbers game—selection pressures and population size

Suggested duration: 100 minutes

INTRODUCTION

A range of selection pressures act on populations of organisms in their natural environments. All living things need resources from their environment and need to get rid of wastes into their environment. Environmental resources are in limited supply and impact on population size. Problems caused by the size of populations usually come to our attention when something appears to be 'out of balance'. Some examples include:

- rabbit, mouse and grasshopper plagues that can cause damage to agricultural productivity
- koala populations becoming too large for the areas set aside for their conservation
- kangaroos requiring culling in areas where their numbers have become too large
- crown-of-thorns sea star invasions on the Great Barrier Reef.

Factors that affect population size are of two kinds:

- factors that increase population size—births (b) and immigration (i) to an area
- factors that decrease population size—deaths (d) and emigration (e) from an area.

MATERIALS

- graph paper
- calculator (optional)

PURPOSE

To consider several sets of population growth figures and make some observations about selection pressures that may affect population size.

Part A—hypothetical population growth

The simplest type of population growth to investigate is a population of bacteria. Bacteria reproduce by a process of binary fission (splitting in two). In ideal conditions, bacteria can reproduce every 30 minutes.

PROCESSING DATA

1 Assuming that there are no factors operating to limit the growth of the population, complete the population data for bacteria in Table 1.

 ISBN 978 1 4886 1931 1

TABLE 1 Bacterial population numbers

Time	Generation number	Number of bacteria
6.30 a.m.	0	1
7.00 a.m.	1	2
7.30 a.m.	2	
8.00 a.m.	3	
8.30 a.m.	4	
9.00 a.m.	5	
9.30 a.m.	6	
10.00 a.m.	7	
10.30 a.m.	8	
11.00 a.m.	9	
11.30 a.m.	10	
12.00 noon	11	
12.30 p.m.	12	
1.00 p.m.	13	
1.30 p.m.	14	
2.00 p.m.	15	

2 Use the following grid to draw a graph representing the data in Table 1. Hint: plot the generation number on the horizontal *x*-axis and number of bacteria on the vertical *y*-axis (1 increment = 2000 bacteria).

ANALYSIS OF RESULTS

3 Describe any problems you encountered when you attempted to draw the graph.

4 Describe the shape of the curve you have drawn.

5 Describe what is happening to the population size as you go from one generation to the next. Outline some of the consequences of this once the population numbers start to become large.

ISBN 978 1 4886 1931 1

6 What name is given to growth of this kind? ______________________

This type of growth is characteristic of a population in which there are no restraints on growth, and so it is not a realistic model. There is one important component missing—the environment and its influence.

Part B—real population growth

In this activity you will plot the data for two natural populations. These populations differ from the hypothetical population because they are continually interacting with their environment.

Population growth of a natural bee colony

Natural beehives are established when a newly emerged queen bee, worker bees (non-reproductive females) and drones (males) fly from the original hive and establish in a new location. A study has documented the number of bees in the hive for 28 weeks from the time of establishment (Table 2).

TABLE 2 Bee colony population growth over 28 weeks after hive establishment

Time since hive establishment (weeks)	Number of bees in beehive
0	1000
2	2500
4	8000
6	22000
8	40000
10	55000
12	72000
14	80000
16	77000
18	80000
20	75000
22	74000
24	78000
26	82000
28	80000

PROCESSING DATA

1 Use the following grid to draw a graph representing the data in Table 2. Hint: plot the time since hive establishment in weeks on the horizontal *x*-axis, and population size on the vertical *y*-axis (1 increment = 10000 bees).

 ISBN 978 1 4886 1931 1

PRACTICAL ACTIVITY 3.1

Population growth of deer on St Paul's Island

Four male and 21 female European reindeer were introduced as an experiment onto St Paul's Island near the coast of Alaska in 1912. The island was free of predators and environmentally favourable to the deer. The population numbers were monitored until 1950 and are provided in Table 3.

TABLE 3 Deer population on St Paul's Island, Alaska (1912–50)

Year	Deer population
1912	25
1915	105
1920	200
1925	315
1930	475
1935	1250
1940	1750
1941	1200
1945	500
1947	300
1948	100
1949	50
1950	8

2 Use the following grid to draw a graph representing the data in Table 3. Think carefully about how best to graduate units on the horizontal and vertical axes.

ANALYSIS OF RESULTS

3 Compare the general shapes of the bee and deer population growth curves. In what ways are the curves:

a similar?

b different?

4 Suggest an explanation for the changing population in the:

a beehive

b deer population

5 In what way is the hypothetical growth curve for bacteria:

a similar to the two natural growth curves (bees and deer)?

b different from the two natural growth curves (bees and deer)?

A number of important points can be made about natural populations:

- The reproductive potential of a population is the maximum number of offspring that are produced.
- A population increases in number exponentially until its demands for resources become equal to the environment's capacity to supply these resources. This is the carrying capacity.
- When population numbers reach this point, more and more organisms die because of the lack of resources. This is environmental resistance.
- Finally, when the number of births plus immigration is approximately equal to the number of deaths plus emigration, the population is in equilibrium.

6 Suggest selection pressures that might contribute to environmental resistance in the:

a bee population

b deer population

CONCLUSION

It has been found that the growth curves for most natural populations are very similar in shape. Four stages have been identified and described for this 'S'-shaped (sigmoid) curve. The stages are identified in the growth curve shown below.

7 Enter each of the following stages of growth into Table 4 so that it corresponds to the correct definition:

exponential　　negative acceleration　　positive acceleration　　equilibrium

TABLE 4 Four stages of growth in natural populations

Name of stage	Description of stage
	• carrying capacity reached • birth rate = death rate • conditions will continue until there is a change in environmental resistance
	• full reproductive potential reached • maximum growth • no environmental resistance
	• as numbers increase, so does environmental resistance, causing either increased death rate, decreased birth rate or both
	• growth increases as reproduction starts • growth is slow at first • no environmental resistance

 ISBN 978 1 4886 1931 1

8 Label the graph by entering the name of the stage and the stage description (Table 4) in the correct place on the growth curve.

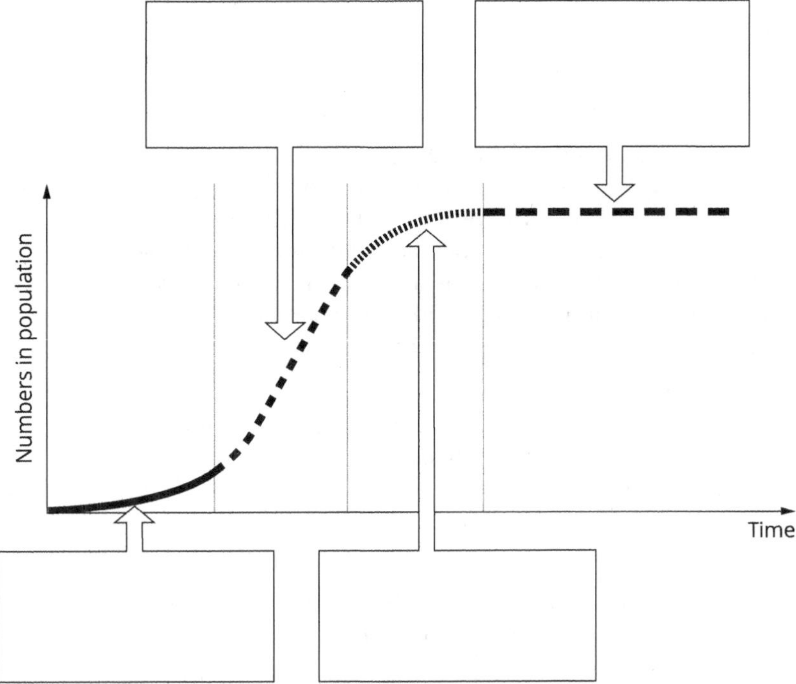

9 Most population data is determined by considering birth rates and death rates. Describe any situations in which the other two factors—immigration and emigration—may operate as a selection pressure on population size.

10 At the start of this activity, several examples of population 'problems' were outlined. Suggest which selection pressures might be responsible for:

a a grasshopper plague. What do you think will eventually become of the grasshopper population?

b the very rapid growth of the crown-of-thorns sea star population. What do you think will eventually become of the sea star population?

11 Consider the examples of population dynamics you studied in this activity. Write a brief summary that describes the pattern of population change over time, and the factors that affect population growth.

RATING MY LEARNING	My understanding improved	Not confident ○ ○ ○ ○ ○ Very confident	I answered questions without help	Not confident ○ ○ ○ ○ ○ Very confident	I corrected my errors without help	Not confident ○ ○ ○ ○ ○ Very confident

ISBN 978 1 4886 1931 1

PRACTICAL ACTIVITY 3.2

Anteater antics—divergent and convergent evolution

Suggested duration: 50 minutes

INTRODUCTION

Have you ever wondered why some apparently unrelated animals show quite remarkable physical likenesses? A bat and a bird, a dolphin and a shark, a sugar glider and a flying squirrel—each pair shares some very obvious structural characteristics. And yet within any one group of animals there can be a remarkable diversity of forms. The most famous example of this was first described by Charles Darwin when he visited the Galapagos Islands in the 1800s. He recorded an astonishing array of beak shapes in the finches native to those islands.

Two evolutionary patterns are operating here. The first is convergence—the evolution of similar structures in organisms that do not share a recent common ancestor. The second is divergence—the evolution of different structures in organisms that share a recent common ancestor.

PURPOSE

To compare and contrast the anatomical features of selected mammals and draw conclusions about the evolutionary relationships between them.

PROCEDURE

Examine the body outlines and read the information related to each of the different anteater species. Anteaters are mammals that feed exclusively on ants and/or termites.

Echidna (*Tachyglossus*)

The Australian echidna is a monotreme (egg-laying) mammal. It has an elongated, beak-like muzzle, with a small mouth at the end. It lacks teeth, but has a long, sticky tongue. It has small eyes and no external ears. The short limbs are equipped with strong claws. The upper surface of its body is covered with heavy spines and it lays leathery, shelled eggs.

Head and body length: 25–53 cm

Body weight: 2.5–6.0 kg

Tachyglossus (Australian echidna)

Aardvark (*Orycteropus*)

The African aardvark is a placental mammal. It has a slender skull with no canine or incisor teeth. Its sticky tongue is extensible and elongated. Its short limbs end in digits with strong claws. It is a lightly-haired animal with long ears and small eyes.

Head and body length: 100–158 cm

Body weight: 50–70 kg

Orycteropus (African aardvark)

ISBN 978 1 4886 1931 1

Giant anteater (*Myrmecophaga*)

The South American giant anteater is a placental mammal. Its skull has an elongated snout and there are no teeth present. The tongue is long and sticky. Its body is covered with long hair. The forelimbs are each equipped with four strong claws.

Head and body length: 100–120 cm

Body weight: 18–23 kg

Myremecophaga (giant South American anteater)

Pangolin (*Manis*)

The scaly pangolin is a placental mammal from Africa and Asia. It has broad, horny scales rather than hairs on its body. It has no teeth and its sticky tongue is over 30 cm long. The pangolin has small eyes and long, powerful claws. It also boasts a large, heavy tail that measures about the length of the rest of the body.

Head and body length: 30–88 cm (not including tail)

Body weight: 4.5–27 kg

Manis (scaly pangolin)

Numbat (*Myrmecobius*)

The Australian marsupial anteater, or numbat, has short hair with a characteristic striped pattern. It has small eyes, about 50 small teeth and a long, sticky, cylindrical tongue. The forelimbs end in powerful claws.

Head and body length: 17–28 cm

Body weight: 275–450 g

Myrmecobius (numbat)

SUMMARY

Use the information about the different anteater species to prepare a report that focuses on the evolutionary relationships between them. Your report should include:

- common features shared by the different species
- differences between the species; include a description of a feature that makes each animal well adapted to its particular lifestyle
- a phylogenetic tree illustrating the evolutionary relationships between the different species; outline the features you used to group as well as distinguish between the different species
- some evidence that suggests divergent evolution from a common ancestor.

Despite belonging to different taxonomic groups, these animals share some characteristics in common. Outline the argument that these animals also represent an example of convergent evolution.

PRACTICAL ACTIVITY 3.2

Summary report

RATING MY LEARNING	My understanding improved	Not confident ◄──► Very confident ○ ○ ○ ○ ○	I answered questions without help	Not confident ◄──► Very confident ○ ○ ○ ○ ○	I corrected my errors without help	Not confident ◄──► Very confident ○ ○ ○ ○ ○

ISBN 978 1 4886 1931 1

PRACTICAL ACTIVITY 3.3

Matching marsupials—similarities and differences between species

Suggested duration: 50 minutes

INTRODUCTION

In this activity you will study the skull structure and dentition of three living marsupials and the same from an unknown extinct animal. Examination of the similarities and differences between them will help you to group them in terms of their evolutionary relationships. Using the technique of comparative anatomy, you will decide to which group the unknown specimen most logically belongs.

PURPOSE

To use anatomical information from the skulls of different but related organisms to draw conclusions about evolutionary relationships.

To use comparative anatomy as a tool in identifying an unknown organism.

BACKGROUND

Australian mammals are uniquely and diversely represented by more than 140 species of marsupials and two species of monotremes. Both groups diversified and flourished on the Australian island continent after it became separated from the supercontinent Gondwana over the period 132-96 million years ago. The infraclass Marsupialia is represented in Australia today by four different Orders:

- Diprotodontia—this group includes 80 different species, all characterised by the presence of two well-developed incisor teeth in the lower jaw; these animals are primarily herbivorous although some include insects and small reptiles in their diets.
- Dasyuromorphia—this group includes 52 species of carnivorous animals; distinguishing features include well-developed canine teeth.
- Peramelemorphia—this group includes eight species of omnivorous animals, that is, they feed on both plant and animal matter, including insects, small birds and mammals as well as grasses and leaves; range of tooth types cater for diverse diet.
- Notoryctemorphia—this order is represented by only one species, the marsupial mole, a sand-burrowing desert-adapted animal that feeds mainly on insect larvae.

Part A—diverse dentition

PROCEDURE

Examine the three figures below showing the skulls and teeth of three mammals. In most mammals the three distinctly different types of teeth we can recognise are the:

- incisors—the chisel-shaped biting teeth located at the front of the mouth
- canines—the pointed, stabbing teeth at the sides of the jaws (most obvious in carnivores)
- grinding teeth—the pre-molars and molars, situated towards the back of the jaws.

The three types of teeth are shown labelled on the drawing of the tiger quoll's skull.

It is important to remember that all three types of teeth may not be present in the jaws of some marsupials.

Tiger quoll (Family Dasyuridae)

The tiger quoll (*Dasyurus maculatus*) is a terrestrial animal and a good climber. It feeds on small birds and mammals as well as reptiles and also includes insects in its diet.

- Head and body length: 40–75 cm
- Body weight: 2.0–3.0 kg

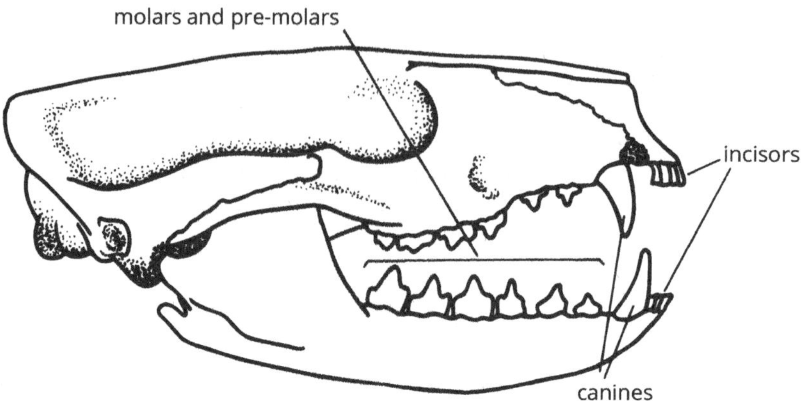

Numbat (Family Myrmecobiidae)

The numbat (*Myrmecobius fasciatus*) feeds exclusively on termites that are collected on its sticky tongue before being swallowed whole.

- Head and body length: 17.5–27.5 cm
- Body weight: 275–450 g

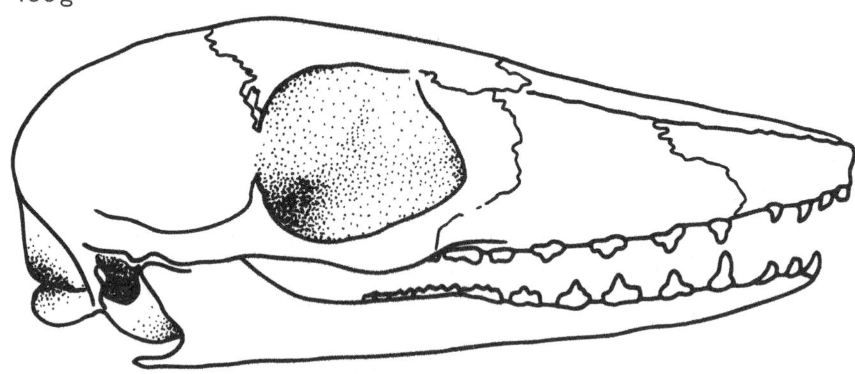

Eastern barred bandicoot (Family Peramelidae)

The eastern barred bandicoot (*Perameles gunnii*) is an omnivorous animal that eats roots, bulbs, insects and occasionally small reptiles.

- Head and body length: 18–45 cm
- Body weight: 550 g

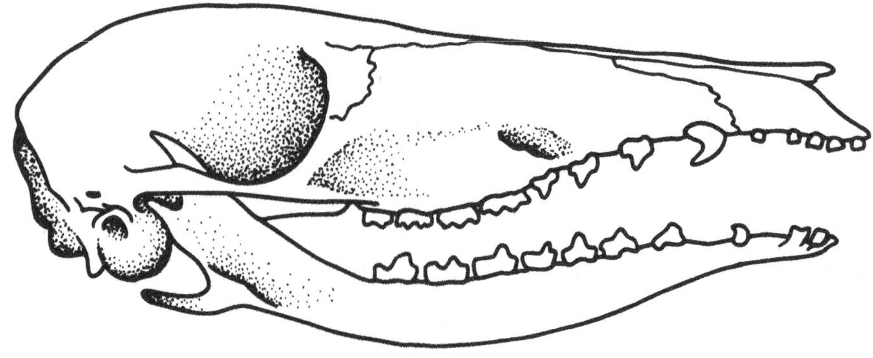

 ISBN 978 1 4886 1931 1

ANALYSIS OF RESULTS

1 All of the animals described so far are marsupials. Describe the characteristics that all marsupials share.

2 There are only four orders of marsupials. Outline a distinguishing feature of each group.

3 Examine the diagrams of each of the skulls presented so far. Use the visual and written information provided about skull shape, dentition and diet to complete Table 1. You also need to decide the Order to which each species belongs. (Note: details for the 'unknown specimen' are to be entered later.)

TABLE 1 Features and classification of marsupials

Animal	Diet	Characteristics of dentition	Order
tiger quoll			
numbat			
bandicoot			
unknown specimen			

4 a Decide which of these species are most closely related. Complete the phylogenetic tree below to show the likely evolutionary relationships between the three species.

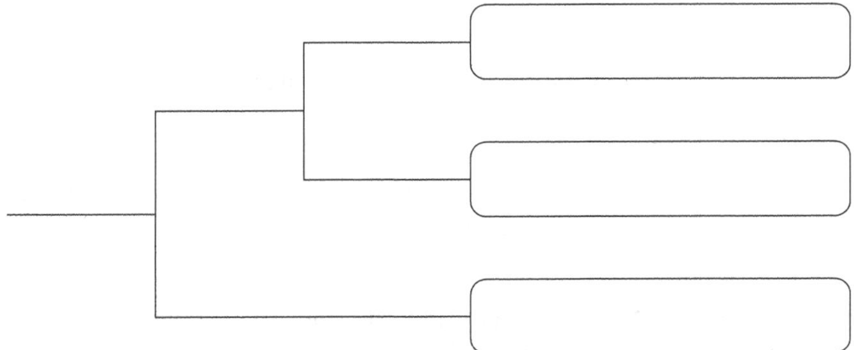

b Outline your reasons for grouping the species as you did.

Part B—comparative anatomy

PROCEDURE

Examine the figure below showing the skull structure and dentition of an unknown specimen.

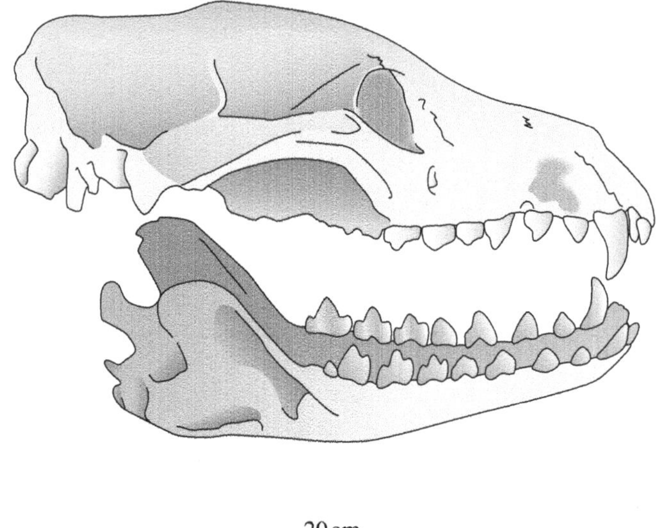

20cm

ANALYSING DATA

1 Looking at the skull in isolation from others in this activity, write out as much information as you can about the animal.

2 Now compare the features of the specimen with each of the known examples in Part A. Which animal does the unknown specimen least resemble? Provide details.

3 With which animal does the specimen appear to share the most similarities? Describe in detail.

4 Enter the details for this specimen into Table 1 on the previous page.

 ISBN 978 1 4886 1931 1

ANALYSIS OF RESULTS

5 Make a suggestion about the identity of the extinct animal represented by the unknown skull specimen, giving reasons for your answer.

CONCLUSION

6 a Outline the basis upon which the different orders of marsupial are grouped.

b There are many different features that could be used as the basis for grouping organisms. Suggest why this approach is logical and useful.

7 Comment on the usefulness of comparative anatomy as a tool in classifying organisms.

RATING MY LEARNING	My understanding improved	Not confident ◄──► Very confident ○ ○ ○ ○ ○	I answered questions without help	Not confident ◄──► Very confident ○ ○ ○ ○ ○	I corrected my errors without help	Not confident ◄──► Very confident ○ ○ ○ ○ ○

ISBN 978 1 4886 1931 1

DEPTH STUDY 3.1

Desirable disadvantages—selection in small populations

Suggested duration: Part A—90 minutes; Part B—120 minutes

INTRODUCTION

This depth study examines secondary-sourced data focusing on a specific example of how one species of organism can affect another to drive change in allele frequencies and the phenotypes in a population. You will process and analyse the data, as well as address a research question. You will communicate your scientific understanding using appropriate terminology.

Part A of this activity raises an interesting conflict about what it means to be biologically fit and demonstrates that this idea needs to be considered in relation to the selection pressures present in the environment of the organisms in question. This activity provides information and data and a short research opportunity, all relevant to the interrelationship between two diseases that affect human populations—malaria and thalassaemia. There is scope to pursue this topic further in Part B.

Part B of this activity provides an opportunity for further investigation on the themes explored in Part A. To prepare yourself for this it will be important to keep a record of issues and questions that the activity raises. One of these will form the basis of your own independent research investigation.

PURPOSE

To investigate the relationship between the incidence of an inherited blood disorder (thalassaemia) and the non-inherited disease, malaria.

To consider the selection pressures acting to maintain normally disadvantageous alleles at relatively high frequencies in populations.

To complete a secondary-sourced investigation into a related theme of your choice based on issues raised in Part A.

Part A—surviving in Sardinia

Thalassaemia

The oxygen-carrying pigment in red blood cells is called haemoglobin. A normal haemoglobin molecule is made up of four small protein chains. There are two alpha (α) chains and two beta (β) chains that fit together into a specific three-dimensional shape that optimises haemoglobin's oxygen-carrying capacity.

In some individuals, the beta haemoglobin molecules are abnormal, resulting in a reduced oxygen-carrying capacity. This condition is called thalassaemia. The shape of the beta haemoglobin molecules is regulated by a gene with two different forms—$\beta+$ is the allele for normal haemoglobin production, and $\beta-$ is the allele for reduced beta haemoglobin production. Most people carry two $\beta+$ alleles and produce normal haemoglobin. If an individual has one normal allele ($\beta+$) and one affected allele ($\beta-$) reduced levels of beta haemoglobin are produced. In this case, distorted but not seriously compromised red blood cells are formed that function fairly normally. When individuals carry this $\beta-$ allele, they are diagnosed with thalassaemia minor and suffer from mild anaemia.

Prior to 1976, no treatment was available for sufferers of thalassaemia major, and death before eight years of age was certain. Today, modern technologies provide treatments that can prevent serious disability and even death.

 ISBN 978 1 4886 1931 1

Malaria and thalassaemia—an unlikely union

Malaria is a non-inherited disease caused by the parasite *Plasmodium falciparum*, which spends part of its life cycle in red blood cells. The transmission agent for malaria is the *Anopheles* mosquito, which must have access to water for much of its life cycle.

Thalassaemia and another blood cell disorder, sickle cell anaemia, are common disorders in Africa, the Mediterranean and parts of Asia. Before 1946 they were widespread, as shown in the first two maps below. Interestingly, these are also areas in which malaria was prevalent at the time.

Several detailed studies have been made of the incidence of malaria and thalassaemia on the Mediterranean island of Sardinia, located off the coast of Italy. The island has swampy coastal plains and a high central mountain range. Many small villages are effectively isolated from one another.

The results of studies considering the incidence of thalassaemia in a number of small villages roughly along a transect line from Oristano to Posada (see map of Sardinia below) are shown in the graph below.

Thalassaemia is a genetic blood disorder that adversely affects an individual's chance of survival, so we would expect the frequency of the alleles responsible for this and other similar conditions to decline within populations. However, the data suggests that this is not the case in Sardinia.

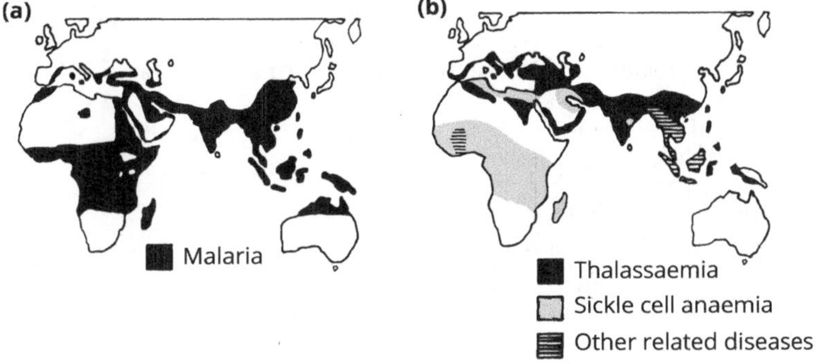

Comparison of (a) the distribution of malaria with (b) the incidence of inherited blood disorders in Africa, the Mediterranean and parts of Asia before 1946.

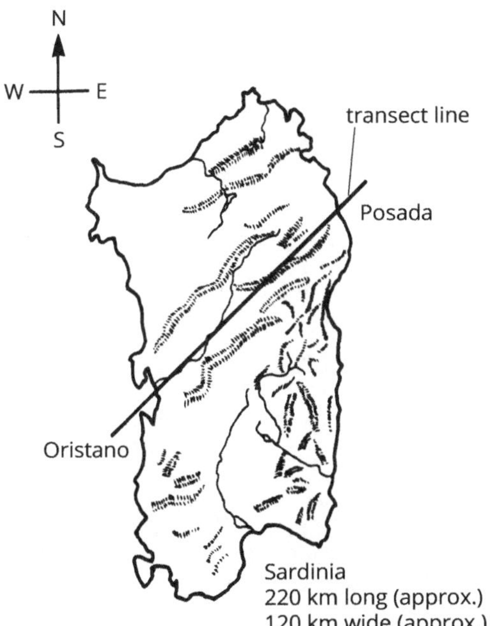

Map of Sardinia showing mountain ranges and approximate position of the transect line

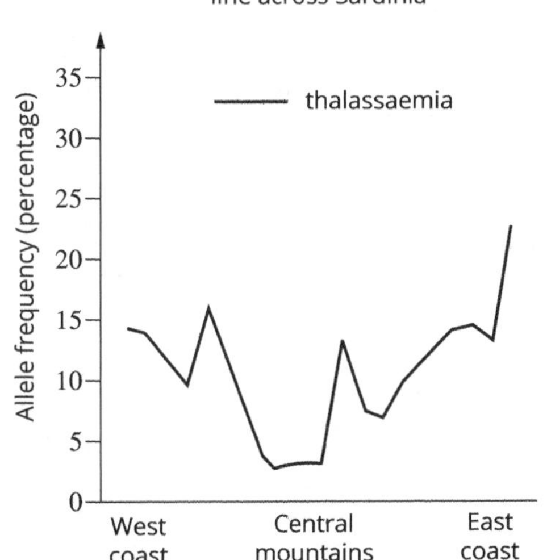

PROCESSING DATA AND INFORMATION

1 a Write down the alleles present in an individual diagnosed with thalassaemia major.

b Suggest how such individuals are likely to be affected.

2 Identify the selection pressure acting on human populations on the island of Sardinia.

3 Suggest a logical explanation to account for the reduction in the incidence of malaria in the regions listed since 1946.

4 Graph the frequency of malaria in relation to altitude using the information in Table 1. Label the axes on your graph.

TABLE 1 Percentage frequency of malaria in Sardinian villages situated at different altitudes

Village	Altitude (m)	Malarial frequency (%)
Benetutti	406	50
Cabras	9	97
Desuio	891	33
Fonni	1000	19
Gatelli	40	96
Gergei	374	60
Isii	523	17
Lanusei	595	25
Lode	345	90
Orosei	19	93
Suni	333	85
Teulada	50	90

 ISBN 978 1 4886 1931 1

ANALYSING DATA AND INFORMATION

5 Describe the relationship between the incidence of malaria and the altitude of populations in Sardinia.

6 Look carefully at the data in the graph on page 139 and in the graph you have drawn. Write a sentence comparing the two sets of data.

7 Write a hypothesis from your observations.

PROCESSING DATA AND INFORMATION

8 Research the life cycle of the malaria *Plasmodium falciparum* to understand how the abnormally-shaped red blood cells of a thalassaemia patient serves as an adaptation that provides a survival advantage in malaria zones. Use the space below to record your findings. A concept map approach will be useful here. Include labelled diagrams where appropriate.

9 Today sufferers of thalassaemia major have improved health and life expectancy compared to the mid-1970s and earlier. Name a management strategy available through modern technology that is responsible for this improvement.

COMMUNICATING

10 Describe how the process of natural selection has led to an increase in the incidence of an otherwise disadvantageous inherited condition, thalassaemia, on the island of Sardinia. Include a discussion of:

- variation
- allele frequencies
- phenotypes that confer a survival advantage given certain environmental pressures
- change in the phenotypic ratio of a population over time.

 ISBN 978 1 4886 1931 1

Part B—issues and questions raised in Part A

This secondary-sourced investigation provides an opportunity for you to further investigate questions and issues raised in Part A. You will communicate your findings in a format of your choice.

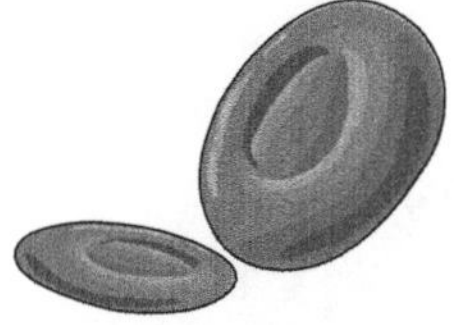

QUESTIONING AND PREDICTING

1 Some ideas for questions or issues that can be investigated are listed below. Add to this list by writing down questions and issues as they arise.

- Emerging technologies in the management of mosquito populations in malaria zones.
- Medical technologies such as gene therapy and stem cell research aimed at effective treatment of thalassaemia.
- Sickle cell anaemia—how can this be a good thing?

2 Consider the various questions and issues you have listed above. Select one of these for further investigation.

Identify your topic:

3 Now that you have selected the theme of your research, it is time to develop the questions you wish to pursue. Think carefully about the aims of your investigation. What is it you want to find out and understand?

PROCESSING DATA AND INFORMATION

4 Refer to the depth study pages of your Toolkit to guide you through the research task. Of particular interest will be pages xiv–xv, which focus on secondary-sourced investigations. The information you gather is important—so too is the way the information is managed and presented in your finished product. Essential aspects to address include:

- locating and recording relevant resources
- keeping your own notes
- including relevant data
- organising the information into useful packets and sorting the order of these
- being mindful of the time required for different parts of the task; set time goals and try to meet them.

A suggested time framework is one hour for research and one hour for the written component.

GO TO ➤ page xiv

COMMUNICATING

5 Prepare your final product. This can be in the form of a journal article, essay, a poster, a digital presentation or any other format agreed upon by you and your teacher. If you choose an article or essay it should be a maximum of 1000 words, not including your references. Your written submission should include:

- a relevant title
- an introduction including a guiding statement about what you have set out to learn (your aims)
- paragraphs that focus on discrete themes
- a concluding paragraph that draws together the themes of the investigation and summarises the information you have collected in the context of your aims
- qualitative data where available.

Remember to:

- communicate your ideas using appropriate terminology
- define new words for your audience
- use footnotes to reference secondary sources such as scientific studies and data
- use a separate reference list prepared according to expected guidelines.

 ISBN 978 1 4886 1931 1

MODULE 3 • REVIEW QUESTIONS

Multiple choice

1 An adaptation is best defined as:

A a structural feature that gives an organism a survival advantage in its particular environment

B a behavioural feature that gives an organism a survival advantage in its particular environment

C a physiological feature that gives an organism a survival advantage in its particular environment

D a feature that gives an organism a survival advantage in its particular environment

2 The cane toad (*Rhinella marina*) was introduced into Australia in 1935 as a biological control mechanism to reduce the population of cane beetles in Queensland sugar cane crops. Unfortunately the cane toad did not successfully reduce cane beetle populations in Queensland. The cane toad preys on insects and snails, as well as small vertebrates such as marsupial mice, birds and their eggs. The cane toad produces a toxin from glands on the skin of its back and shoulders that is poisonous to native carnivores that might prey on it (for example, goannas and quolls). The cane toad represents a selection pressure that acts to:

A reduce cane beetle populations

B reduce quoll populations

C increase goanna populations

D compete with marsupial mice

3 The theory of evolution by natural selection:

A was developed by English naturalist Charles Darwin in the 1800s

B was developed by English naturalist Alfred Wallace in the 1800s

C is widely recognised as the mechanism of evolution of life on Earth

D all of the above

4 The fossil record demonstrates:

A an increase in the number of organisms over small expanses of geological time

B a decrease in the diversity of organisms over large expanses of geological time

C an increase in the number and diversity of species over large expanses of geological time

D an increase in the diversity of species over small expanses of geological time

5 Select the phylogenetic tree that best summarises the evolutionary relationship between sharks, dolphins and whales.

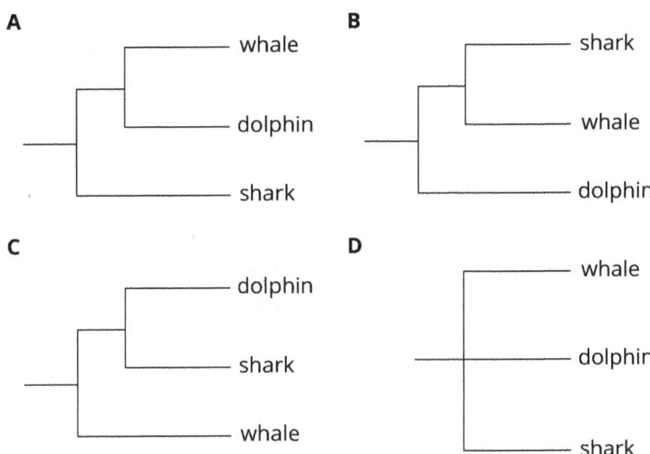

6 The relatively rapid speciation of 13 different species of finches on the Galapagos Islands from a single ancestral species is an example of:

A microevolution

B macroevolution

C adaptive radiation

D instant speciation by polyploidy

Short answer

7 Cuttings through a rock face in a quarry revealed a series of fossils in the rock strata (see site 1 in the diagram below). Road cuttings at other locations are represented in the diagram as sites 2 and 3. Assume that the layers of rock were deposited in the order in which they appear.

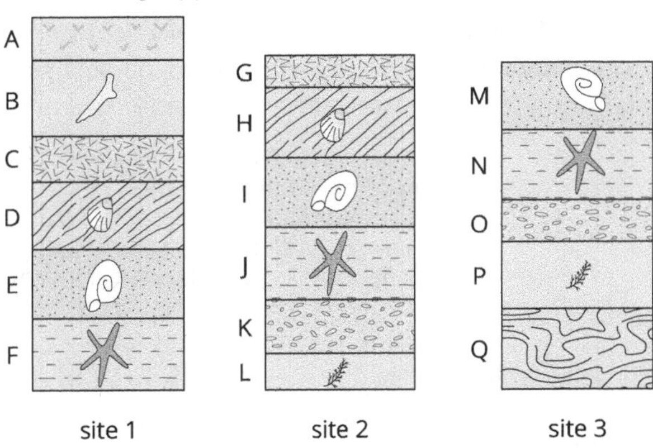

a i Define the term 'fossil'.

ii Describe two different ways in which organisms might be fossilised.

b Examine the diagram and identify three rock strata that appear to have been deposited at about the same time. Give reasons for your answer.

c Identify the rock stratum that contains the:

i oldest fossils

ii youngest fossils

d Suggest two reasons to account for the incomplete nature of the fossil record.

8 Examine the illustration representing embryological development in a group of vertebrates.

fertilised egg

early cell division

A B C D

late fetal

a Explain what is meant by comparative embryology.

b State a conclusion that can be drawn about the evolutionary relationship between the vertebrates represented. Explain your reasoning.

9 Molecular homology is a tool used by scientists to help establish evolutionary relationships between organisms. The diagram below represents the amino acid sequences for five mammals. Differences in amino acid sequences between the organisms are highlighted.

```
human KKASKPKKAASKAPTKKPKATPVKKAKKKLAATPKKAKKPKTVKAKPVKASKPKKAKPVK
mouse KKAAKPKKAASKAPSKKPKATPVKKAKKKPAATPKKAKKPKVVKVKPVKASKPKKAKTVK
  rat KKAAKPKKAASKAPSKKPKATPVKKAKKKPAATPKKAKKPKIVKVKPVKASKPKKAKPVK
  cow KKAAKPKKAASKAPSKKPKATPVKKAKKKPAATPKKTKKPKTVKAKPVKASKPKKTKPVK
chimp KKASKPKKAASKAPTKKPKATPVKKAKKKLAATPKKAKKPKTVKAKPVKASKPKKAKPVK
      *** ********** ************** ****** **** ** ********** * **
```

a Explain how molecular homology is used to understand the evolutionary relationship between organisms.

b Use the information about amino acid sequences to draw a phylogenetic tree that represents the likely evolutionary relationship between the organisms.

c Explain your reasoning in grouping the organisms the way you have in question 9b.

 ISBN 978 1 4886 1931 1

Extended response

10 Biogeography examines the geographical distribution of organisms.

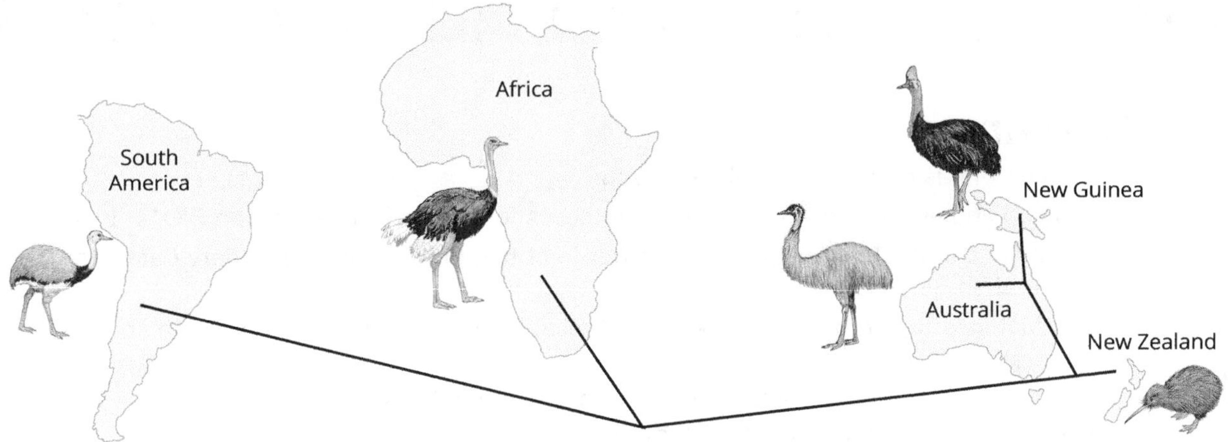

a Describe two pieces of evidence that suggest the different species of flightless birds share a common ancestor.

b Draw a flowchart to provide a detailed account of the key steps in the speciation of flightless birds on the different continents. Include the following terms in your answer:

speciation	selection pressures	supercontinent	natural selection
allopatric	common ancestor	continental drift	reproductive isolation

MODULE 4 Ecosystem dynamics

Outcomes

By the end of this module you will be able to:

- develop and evaluate questions and hypotheses for scientific investigation BIO11-1
- design and evaluate investigations in order to obtain primary and secondary data and information BIO11-2
- conduct investigations to collect valid and reliable primary and secondary data and information BIO11-3
- select and process appropriate qualitative and quantitative data and information using a range of appropriate media BIO11-4
- analyse and evaluate primary and secondary data and information BIO11-5
- analyse ecosystem dynamics and the interrelationships of organisms within the ecosystem BIO11-11

Content

POPULATION DYNAMICS

INQUIRY QUESTION **What effect can one species have on the other species in a community?**

By the end of this module you will be able to:

- investigate and determine relationships between biotic and abiotic factors in an ecosystem, including: (ACSBL019) S CCT ICT PSC
  - the impact of abiotic factors (ACSBL021, ACSBL022, ACSBL025)
  - the impact of biotic factors, including predation, competition and symbiotic relationships (ACSBL024)
  - the ecological niches occupied by species (ACSBL023)
  - predicting consequences for populations in ecosystems due to predation, competition, symbiosis and disease (ACSBL019, ACSBL020) S
  - measuring populations of organisms using sampling techniques (ACSBL003, ACSBL015) ICT N
- explain a recent extinction event (ACSBL024) S L

PAST ECOSYSTEMS

INQUIRY QUESTION **How do selection pressures within an ecosystem influence evolutionary change?**

By the end of this module you will be able to:

- analyse palaeontological and geological evidence that can be used to provide evidence for past changes in ecosystems, including but not limited to: CCT ICT
  - Aboriginal rock paintings AHC
  - rock structure and formation
  - ice core drilling

- investigate and analyse past and present technologies that have been used to determine evidence for past changes, for example: (ACSBL005)
  - radiometric dating
  - gas analysis
- analyse evidence that present-day organisms have evolved from organisms in the past by examining and interpreting a range of secondary sources to evaluate processes, claims and conclusions relating to the evolution of organisms in Australia, for example: (ACSBL005, ACSBL027) CCT ICT
  - small mammals
  - sclerophyll plants
- investigate the reasons for changes in past ecosystems, by:
  - interpreting a range of secondary sources to develop an understanding of the changes in biotic and abiotic factors over short and long periods of time (ACSBL025, ACSBL026) ICT
  - evaluating hypotheses that account for identified trends (ACSBL001) CCT

FUTURE ECOSYSTEMS

INQUIRY QUESTION **How can human activity impact an ecosystem?**

By the end of this module you will be able to:

- investigate changes in past ecosystems that may inform our approach to the management of future ecosystems, including:
  - the role of human-induced selection pressures on the extinction of species (ACSBL005, ACSBL028, ACSBL095) AHC S
  - models that humans can use to predict future impacts on biodiversity (ACSBL029, ACSBL071) AHC S N
  - the role of changing climate on ecosystems S
- investigate practices used to restore damaged ecosystems, Country or Place, for example: AHC S
  - mining sites
  - land degradation from agricultural practices

Key knowledge

Population dynamics

RELATIONSHIPS BETWEEN BIOTIC AND ABIOTIC FACTORS IN AN ECOSYSTEM

An **ecosystem** is a system of organisms interacting with one another and with their non-living surroundings. A **community** is a group of different species living together and interacting with one another in a particular **habitat**. An organism's habitat is the place where it lives at a particular time. A **microhabitat** is a small living space within a larger habitat.

An organism's environment is the sum total of all of the factors that affect it during its lifetime.

Abiotic factors are non-living or physical factors in the environment. **Biotic factors** are living organisms in the environment.

A self-sustaining ecosystem does not generally require external input to preserve the species within it, with the exception of sunlight. A lake aquatic system is self-sustaining because all the species within the community can survive and reproduce without interaction with factors outside the lake. A cliff-side nesting ground of gulls is not self-sustaining because the gulls need to go out to sea to catch fish.

Abiotic factors shape ecosystems

Abiotic factors shape the kind of ecosystem that develops in the first place (Table 4.1). Climate is a significant factor. For example, ecosystems in polar zones are fundamentally different from those in temperate and tropical zones. Other factors include altitude, temperature, geology and light intensity. Each of these factors influence the **distribution** and **abundance** of organisms.

Biogeography is the study of the distribution of organisms. The distribution of organisms depends on the kind of environment in question. Organisms must be adapted to particular environments to survive there.

Biodiversity refers to the range of different kinds of organisms on Earth. Biodiversity is higher in geographical regions closer to the equator and decreases closer to the poles.

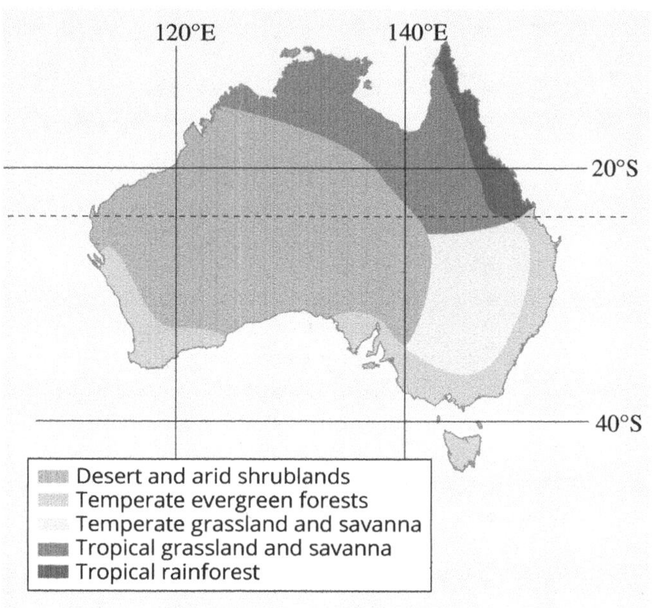

FIGURE 4.1 Distribution of major biomes of Australia

A **biome** is a very large area characterised by particular environmental conditions. For example, desert, tundra, alpine and forest constitute different biomes (Figure 4.1). The **biosphere** is the sum of all the ecosystems on Earth.

The three main ways of naming an ecosystem are:

- focusing on major abiotic factors, for example marine, freshwater or terrestrial. Such factors determine the kind of plants and animals that can live there
- identifying the dominant species or most abundant species of the system, for example oyster community or eucalyptus forest
- describing it according to the plant community of the system. The description is made up of a combination of the tallest or dominant plant and the percentage of sunlight coverage of the canopy, for example closed forest.

TABLE 4.1 Abiotic characteristics of habitats

Terrestrial			Aquatic	
Tropical	**Temperate**	**Semi-arid to arid desert**	**Freshwater**	**Marine**
• high rainfall • warm • rainforests	• cool • high altitudes • winter snow	• low rainfall • desert-like environment • hot	• relatively low water salt concentration; inland lakes, rivers, streams	• relatively high water salt concentrations; sea/ocean

 ISBN 978 1 4886 1931 1

Biotic factors shape ecosystems

The abiotic factors in a particular area are key factors in determining the kind of ecosystems that develop and the kinds of organisms that can live there. Organisms are adapted to the environments in which they live. They interact with one another and with their non-living surroundings. Their survival depends on these interactions.

Organisms living together in an environment constantly interact with and influence each other. These interactions can be benign, beneficial or harmful to the organisms.

Competition

Organisms compete for resources. When organisms have the same survival requirements, they compete for the limited resources. **Competition** can be interspecific or intraspecific.

In **interspecific** competition organisms of different species compete for the same resources, for example various bird species in a forest compete for nest sites.

In **intraspecific** competition different individuals of the same species compete for resources, for example male koalas fight for mating rights with females.

Competitive exclusion occurs when one species successfully utilises a resource at the exclusion of another—they cannot both use the resource.

Predation occurs when one organism hunts and eats another.

Symbiosis is a relationship in which two organisms live in close association with each other. Usually at least one organism benefits from the relationship (Figure 4.2). Table 4.2 summarises different kinds of symbiotic relationships.

FIGURE 4.2 The ivy vine growing on the trunk of this tree is an epiphyte. It is supported by the tree but does not obtain food from it, and it does not affect the health of the tree. This interaction is a type of symbiosis.

TABLE 4.2 Types of symbiosis

Type	Species A	Species B	Examples
Amensalism	no effect	harmed	• The human activity of deforestation is detrimental to the species of plants and animals that inhabit the forest, however there is no biological benefit to the humans.
Mutualism	benefits	benefits	• Algae and fungus live together as lichen. • Plant roots and soil fungi co-exist, exchanging sugars and phosphorus in a relationship specifically called mycorrhiza. • There are two kinds of mutualism. In obligate mutualism both species in the relationship are completely dependent on one another. Lichens are an example of obligate mutualism. In facultative mutualism both species in the relationship benefit from the relationship but it isn't essential to their survival. Birds perched on the backs of cattle feeding on ticks is an example—the bird benefits by accessing food and the cattle benefit by the removal of ticks and other ectoparasites.
Commensalism	benefits	no effect	• Epiphytes (perching plants), use the tree as a surface to live on for access to sunlight. • Remora live on the sides of sharks, eating food scraps from the sharks kills.
Parasitism	benefits	harmed	• Heartworm feed on the blood of dogs. • Mistletoe absorbs nutrients from a host tree.

Feeding interdependencies

Included in the wide range of interactions between organisms in ecosystems are the feeding relationships between species.

Food chains are linear feeding relationships. Each food chain begins with a **producer** organism (Table 4.3). The direction of arrows indicates the flow of energy along a food chain. Energy initially harnessed by producers in photosynthesis is passed along food chains from first-order to higher order **consumers** (Figure 4.3). The first consumer in the food chain is called the first-order consumer; the second consumer is the second-order consumer, and so on.

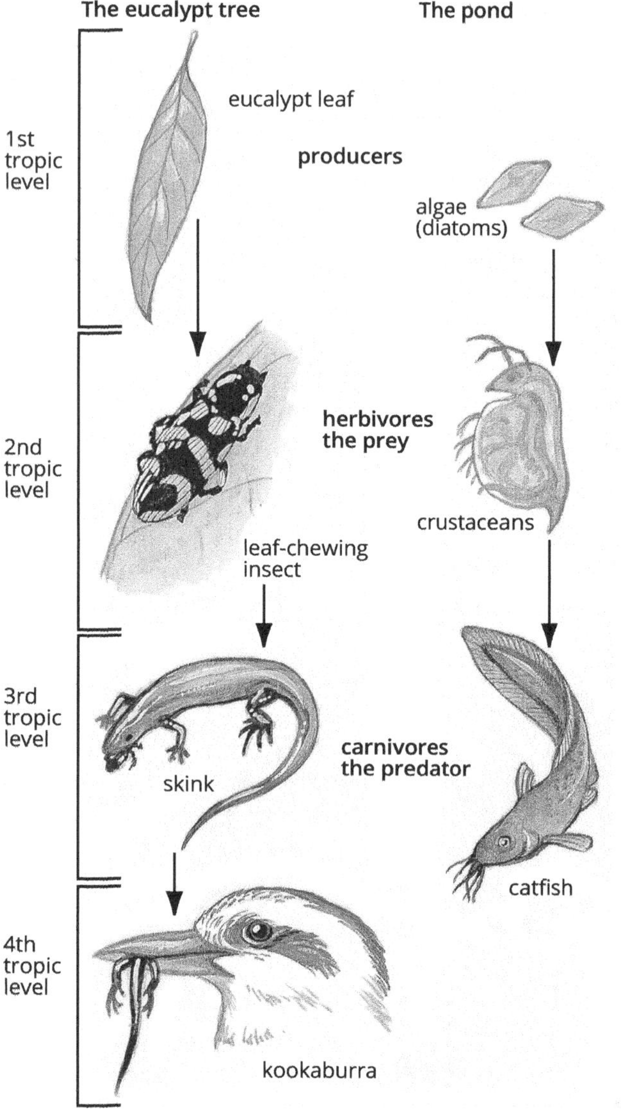

FIGURE 4.3 Two simple food chains, one in the eucalypt tree and one in the pond. These include examples of predator–prey food chains.

Producers are typically plants and are always autotrophic organisms, manufacturing their own organic compounds from inorganic ones. Consumers are heterotrophs—they do not manufacture their own organic compounds and must consume other organisms to acquire these. Consumers can be herbivorous (consume vegetation), carnivorous (consume the flesh of other organisms) or omnivorous (consume both plant and animal matter).

A **trophic level** in the scheme of feeding relationships refers to the position an organism is placed in the food chain. The first trophic level of food chains and **food webs** contains most of the energy. There is less energy in subsequent trophic levels.

TABLE 4.3 Components of food chains

Type	Acquires energy through	Example
Producer/ autotroph	photosynthesis	all green plants
Consumer/ heterotroph	feeds on other organisms	all animals
• herbivore	feeds on plant material	kangaroo
• carnivore	feeds on other animals	killer whale
• scavenger	feeds on dead animals	Tasmanian devil
• detritivore	feeds on decomposing and waste material	worm
• omnivore	feeds on plant and animal matter	bear
Parasite	takes nutrients from other organisms often harming the host	tapeworm, mistletoe, flea
Decomposer	breaks down dead organic matter	bacteria

Predator–prey food chains are linear sequences of 'who eats who'.

In **parasite–host** food chains:

- a smaller organism (the parasite) lives on or inside a larger organism (the host)
- the parasite takes nutrients and resources directly from the body of its host
- the parasite has a complex life cycle, often involved in several different food chains at different stages of its development.

In **detrivore–decomposer** food chains detrivores break down dead organisms into smaller particles, which are then further broken down by decomposers.

A food web is a network of interconnected food chains. Food webs show how various organisms are a part of many linking food chains within a community. A particular organism can be at multiple trophic levels within a food web. Removing a species from a food web affects other organisms in the food web.

Ecosystems and food webs include **keystone species**. Keystone species play a critical role in maintaining the integrity of an ecosystem. Their removal destabilises the ecosystem (Figure 4.4).

 ISBN 978 1 4886 1931 1

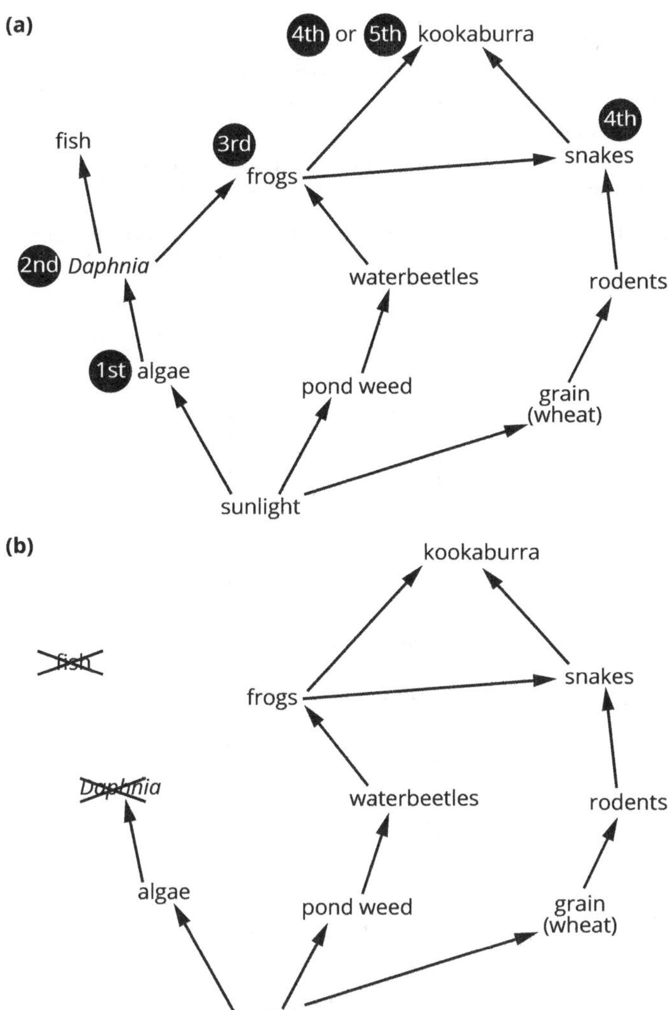

FIGURE 4.4 (a) A hypothetical food web. The kookaburra is at both the 4th and 5th trophic levels, i.e. it is both a 4th and a 5th order consumer. (b) The domino effect on trophic levels—when *Daphnia* are removed from the food web, frogs lose a source of food but survive, while fish lose their food source and die out. Complex webs are more stable because organisms are involved in more than one food chain.

ECOLOGICAL NICHES

All organisms in an ecosystem interact with other organisms, and with their non-living surroundings. These relationships are important in maintaining the stability of the ecosystem. The role that a particular species occupies in an ecosystem is called its **ecological niche**. For example, the niche of plants in an ecosystem is that of producer—through photosynthesis they manufacture organic compounds for their own growth. In so doing they provide an energy source for consumers. Bacteria occupy a decomposer niche, recycling materials from organisms that have died so they become available to be used again.

The competitive exclusion principle states that two species cannot have the same niche in an ecosystem. This can be avoided by **resource partitioning** (different species using available resources differently; this avoids competition).

PREDICTING AND MEASURING POPULATION DYNAMICS

Factors affecting abundance and distribution

The distribution and abundance of species is related to the availability of their requirements (Table 4.4). This includes food, water, shelter, mates and suitable environmental conditions. Environmental conditions must provide conditions within which organisms can survive. This is called the tolerance range for the organism.

Factors such as competition for resources, disease and chance environmental events also impact on species and their **population** sizes in ecosystems. Such factors can be described as density-independent or density-dependent.

Density-independent factors impact on populations regardless of the population size or density. Examples include natural disasters such as bushfire, cyclone, flood and drought.

Density-dependent factors are those that influence population size and density, but the magnitude of their effect depends on the existing population density. Examples include competition, predation, crowding, parasitism and disease.

Anthropogenic factors (anthro: human) are human-induced factors that influence ecosystems. Examples include pollution, deforestation, construction (buildings, cities, roads) and noise.

Determining population sizes

Various factors affect the population size of organisms in ecosystems. Ecologists must take all of these into account when monitoring population abundance. This includes birth and immigration (individuals moving into an existing population) and death and emigration (individuals moving out of an existing population).

Populations increase when the combined rates of birth and immigration are greater than the combined rates of death and emigration.

Rule: abundance = birth + immigration − death − emigration

- Limiting factors such as the presence of predators restrict populations of organisms from exponential growth, thereby keeping populations in check.
- Population explosions (exponential growth) occur when an ecosystem doesn't have sufficient factors to limit the population growth. For example, the crown-of-thorns sea star in the Great Barrier Reef has abundant food resources, high reproductive success and few predators.
- A population crash occurs when the number of individuals exceeds the carrying capacity of the ecosystem. For example, previously abundant resources such as food become scarce due to overgrazing, resulting in a rapid decline in population.
- When factors such as birth rate and immigration are in balance with death rate and emigration, population size is maintained at a constant level.

ISBN 978 1 4886 1931 1

TABLE 4.4 Factors that determine distribution and abundance of organisms

	Abiotic characteristics of environment	Biotic characteristics of organism	Biotic interactions between organisms
Explanation	the physical limitations of the environment	the physical, functional and behavioural adaptations of the organism	how all the species in a community affect and influence each other
Example	the availability of water in a desert	koalas' ability to digest leaves that other herbivores cannot digest	prey populations will explode without predators to keep them in check

Measuring populations in an ecosystem

Point sampling

Point sampling involves population estimation by counting individuals at selected points. The sample points may be random or predetermined. While point sampling is useful for broadly establishing the kinds of organisms present, it also has the potential of missing some species.

Quadrats and transects

A **quadrat** is an area marked out with a frame for the purpose of gathering data related to populations of organisms in a given area (Figure 4.5).

- It is usually 1 m^2 but can be adapted to suit the specific ecosystem.
- Organisms inside the quadrat are counted and recorded.
- A number of quadrats placed randomly in the habitat can provide a useful estimate of the presence, abundance and density of different species within the area.

A **transect** is a line marked out randomly through a habitat.

- Every organism on the line at regular intervals or within the transect is recorded.
- Variations in community composition throughout the habitat can be assessed.
- Line transects are time-efficient and can minimise disturbance to the environment. However, species of low abundance can be missed.
- Belt transects extend out a specific distance to either side of the line. They are time-intensive but can provide more accurate estimates of community populations.

Permanent quadrats and transects can be used to measure, estimate and predict changes in the diversity and abundance of populations over time.

Mark-recapture

Mark-recapture is a method used to monitor animal populations. Individuals are tagged and released—they can then easily be identified when they are spotted again. This avoids counting individuals multiple times or missing them altogether.

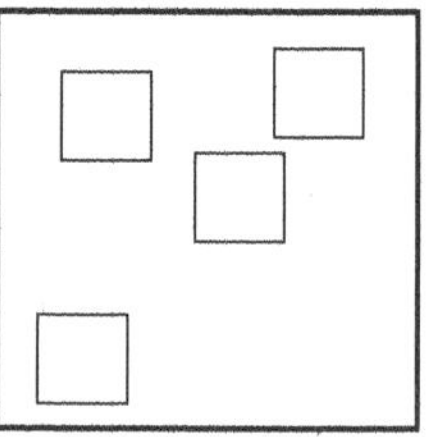
quadrat sampling

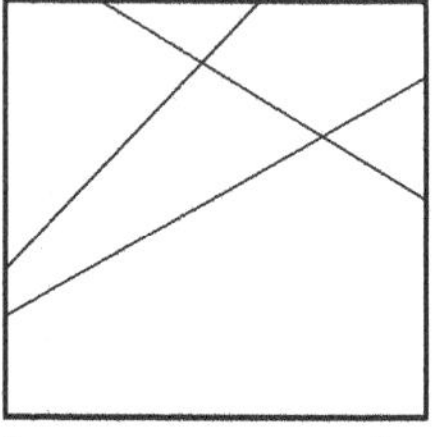
line transect sampling

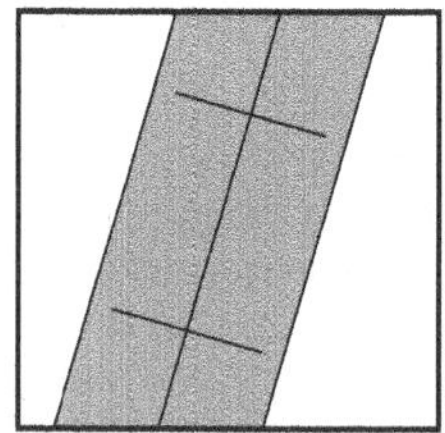
belt transect sampling

FIGURE 4.5 Sampling methods

Monitoring environmental factors

Environmental factors affect the ability of organisms to function at an optimal level. Various tools and techniques are used to monitor environmental factors (Table 4.5). The use of such tools and techniques is important in the management of plants, animals and environmental resources.

ISBN 978 1 4886 1931 1

TABLE 4.5 Techniques to monitor environmental factors

Factor	Technique for monitoring factor	Notes
pH	A pH probe measures the degree of acidity/alkalinity in soil or water.	Different species of plants have particular pH requirements in order to grow efficiently. Australian inland waters have a pH range of 6 to 9. Aquatic organisms tolerate a wide range of pH, but will be adversely affected if pH goes beyond this range.
Light	A light meter measures the intensity of light at different levels of penetration in a habitat.	The intensity of light is an important factor affecting the distribution of producer organisms (plants) in a given habitat. This, in turn, affects the abundance and distribution of animals that feed either directly or indirectly on the plants.
Oxygen	Probes are available to monitor oxygen concentration in a habitat.	Oxygen is a requirement for organisms; when oxygen availability is reduced below the tolerance limit for organisms, cellular respiration cannot occur at a rate sufficient to provide energy to cells and the organism dies.
Carbon dioxide	Probes are available to monitor carbon dioxide concentration in a habitat.	Plants require carbon dioxide for photosynthesis; high levels of this gas are toxic to animals.

EXTINCTION

Species become **extinct** as a result of environmental change. When environmental change is extreme there may be no individuals in the population that are sufficiently adapted to survive, resulting in the population becoming extinct. Examples of environmental change include the introduction of disease, competitors or predators. Climate change (for example, ice ages) and habitat loss are other examples.

Over millions of years of geological time, species evolve and species become extinct. This is called background extinction. In some circumstances a large number of species become extinct in a relatively short time or in response to a specific environmental factor. Such extinction events are called mass extinctions. For example, changes to global climate mean some species cannot survive. A meteorite strike around 65 million years ago is believed to be responsible for significant climate and ecosystem changes that contributed to the extinction of the dinosaurs. Human intervention, by way of hunting, is largely responsible for the extinction of the Australian thylacine (Tasmanian tiger) only last century.

Past ecosystems

ECOSYSTEM DYNAMICS: CHANGES AND CAUSES

Ecosystems are dynamic, undergoing constant change. Often change is slow and not immediately obvious. Other changes are dramatic, resulting in rapid, significant outcomes.

Factors in ecosystems that cause change are called disturbances. When change is slow and fluctuating in a relatively stable cycle, ecosystems tend to remain relatively stable. **Feedback loops** help to regulate this stability. Positive feedback occurs when a disturbance factor increases a component of an ecosystem it acts upon. For example, when food is abundant, populations increase due to increased birth rate and immigration. However, the increased population can lead to overexploitation of the limited food resource, leading to a reduction in the population. This is negative feedback.

The way ecosystems are affected by disturbances depends upon their intensity and frequency. Balanced ecosystems are resilient to low-level disturbance and stability prevails.

The **Intermediate Disturbance Hypothesis** states that the biodiversity in an ecosystem is maximised when there is an intermediate level of disturbance. High and low-level disturbances reduce species diversity.

Changes to the Great Barrier Reef provide an example of high-level disturbance in a unique ecosystem. The introduction of the predatory crown-of-thorns sea star represents a high level of disturbance leading to significant declines in coral populations. Increasing global temperatures are also responsible for reef damage. Higher water temperatures cause stress to symbiotic algae that live in coral polyps; they leave the corals' tissue, which then causes the bleaching effect. Acidification of seawater caused by higher atmospheric carbon dioxide (greenhouse gas) further contributes to stress on corals. The increased intensity and frequency of extreme weather events such as cyclones is another disturbance factor. All of these factors have contributed to destabilising the Great Barrier Reef ecosystem.

TECHNOLOGY AND EVIDENCE FOR PAST ECOSYSTEM CHANGE

It is more difficult to assess changes in past ecosystems. Evidence of ecological change is available through technologies such as radiometric dating, drilling of ice cores and ancient human records.

Radiometric dating uses the rate of radioactive decay of isotopes in the fossilised remains of organisms to estimate how long ago they died. This provides a picture of change over geological time, as it demonstrates a change in the kinds of organisms that lived in the past, particularly those that are now extinct.

Australian megafauna are an example of ecological change. Radiometric dating of the fossils of megafauna shows that they roamed the Australian landscape between 10 million and 10 000 years ago.

Ice cores represent another kind of technology that provides information about abiotic factors at different times in the past. Air bubbles trapped in ice contain concentrations of atmospheric gases that existed at the time of freezing, providing a snapshot of the atmosphere literally frozen in time. Analysis of Antarctic ice cores demonstrates the changing composition of atmospheric gases, such as carbon dioxide, over the last 800 000 years. Interpretation of such information means understanding more about related factors such as atmospheric temperature. Patterns of change are established that help us understand ecosystem dynamics over time.

Human records also provide information about the nature of ecosystems in the past. For example, Indigenous Australians have lived on the Australian continent for over 60 000 years and have left evidence in different forms that tell us about their environment. **Rock paintings** depict the presence and distribution of now extinct megafauna. **Middens** (shell and bone debris left behind from meals) inform us about Indigenous culture, including the food they ate and how meals were prepared.

LIVING EVIDENCE OF ECOSYSTEM CHANGE

Until around 30 million years ago, the Australian continent was still attached to Antarctica. Originally both continents were part of the southern supercontinent of Gondwana, which was composed of the present-day landmasses of New Zealand, Antarctica, India, South America, Africa and Madagascar. The supercontinent was located much further south, so environmental conditions such as climate were very different.

In the 30 million years since Australia completed its separation from Antarctica, it has continued its northward movement to its present position. This has been accompanied by changes in abiotic factors such as climate, which has precipitated changes in the kinds of organisms inhabiting the land. Many plants and animals adapted to the colder more southern climate have become extinct. Other species have evolved adaptations that make them well suited to the warmer, drier conditions that prevail over most of the continent.

Sclerophyll plants are typically characterised by hard leaves that can withstand the hot, dry conditions that prevail in inland Australia. They are also adapted to low fertility soils. Sclerophyll plants such as *Eucalyptus* and *Acacia* did not exist before Australia's separation from Gondwana. They are recent in geological time, appearing between 50 and 35 million years ago.

Small mammals are another example of evolutionary change in the changing Australian landscape. The large body size and small surface-area-to-volume ratio of the megafauna made them well adapted to glacial periods during the ice ages. As Australia drifted further north and conditions became hotter and drier, being small offered greater survival advantages than being large (small animals can lose heat more efficiently than large animals). The last of the megafauna are believed to have become extinct around 10 000 years ago. They have given way to smaller, more adapted animals, such as present-day koalas, platypus and kangaroos.

Such species change is evidence of changing selection pressures in ecosystems. Some species become extinct while others evolve. We continue to observe changing selection pressures that impact on species. Human activity is a key factor. Clearing forests and natural grasslands for human habitation, agriculture and roads means fragmented habitats for some species. Such **habitat fragmentation** isolates populations, reducing or stopping reproduction between members of different populations. This results in reduced diversity in populations, leaving them more vulnerable to selection pressures, particularly during periods of environmental change. Extinction can follow if the species is not sufficiently adapted to the new environmental conditions and selection pressures.

Invasive species represent another example of a factor that places selection pressure on organisms. Invasive species impact populations in a variety of ways. Cane toads introduced into Australia outcompete native species for resources, disrupt nesting sites of some native birds and poison native carnivores that prey on them. This selection pressure has pushed the northern quoll to the Critically Endangered list. Rabbits and prickly pear are other invasive species that have been introduced to Australia and place selection pressures on native plants and animals.

Biosecurity is the practice of preventing non-native species of plants and animals from entering Australia. This is aimed at eliminating potentially invasive species from seriously impacting Australian ecosystems, altering them and threatening native species.

Future ecosystems

Our growing understanding of the causes of change in ecosystems, and particularly the ways in which humans have facilitated change in recent centuries, is useful in predicting impacts on biodiversity and planning future strategies to maintain natural ecosystems where possible and to minimise impact in others.

HUMAN-INDUCED CHANGES

Human-induced changes to native ecosystems such as habitat destruction, habitat fragmentation and the introduction of invasive species have potentially far-reaching consequences, including the extinction of species. Invasive species impact on native species through predation, habitat modification and competition. Successful invaders typically have few enemies, are associated with humans, show rapid growth, maturation and reproduction, and easily adapt to new environments. Pollution is another key factor that has resulted in large-scale impacts on global ecosystems, such as climate change. Overexploitation also causes many species to become extinct.

Biodiversity hotspots are areas that have been identified as having high biodiversity and high habitat destruction. Biodiversity hotspots require special attention to ensure vulnerable species are protected. This involves management of the ecosystem to reduce habitat destruction and restore the habitat.

ISBN 978 1 4886 1931 1

Vulnerable ecosystems show some common features. These include:

- disturbed or degraded ecosystems
- high resource availability that allows invasive species to thrive
- moderate climate, making it accessible to more species.

PREDICTING IMPACTS ON BIODIVERSITY

Predicting environmental changes and their impacts on biodiversity can allow scientists to take action to prevent population declines and extinctions. Some methods of monitoring and predicting changes are bioindicators, Representative Concentration Pathways (RCPs) and species distribution models.

- **Bioindictors** are species that show particular sensitivity to environmental change. They are useful in monitoring the health of ecosystems.
- **Representative Concentration Pathways (RCPs)** use information about population growth, land use, technology and energy generation to predict the trajectory of greenhouse gas emissions. They are a critical aspect of strategic management of greenhouse gas emissions.
- **Species distribution models** use data about the distribution of species and their ecological niche requirements to predict locations outside of their natural range where they might survive. This serves as an important **conservation** strategy to keep vulnerable species from extinction.

MANAGING AND CONSERVING BIODIVERSITY

The last 200 years has seen a dramatic decrease in biodiversity on Earth. This has been largely due to human activity, either by direct or indirect means. Today, strategies are in place to maintain biodiversity and to reduce the rate of biodiversity loss. As we understand more about the complex relationships between organisms and between organisms and their environment, as well as the impacts of human-induced change in ecosystems, we appreciate more and more the value of biodiversity.

We identify the value of biodiversity in different ways. These include aesthetic value, practical value and ecological value (Table 4.6).

TABLE 4.6 Reasons for species conservation

Aesthetic	Ecological	Practical
The simple pleasure of enjoying the living record of biodiversity.	All species interact with one another, including us. All organisms are linked and the loss of one could affect many others.	Plants provide us with food, shelter, fuel, medicines, etc. Animals provide us with food, clothing, labour, companionship, etc.

Species also have intrinsic value—they have value in their own right. Social and cultural value refers to the value societies attach to different species because of the place they have in cultural identity and traditional expressions.

We are also aware of the value species offer our society in terms of the potential for medicinal applications, such as in the manufacture of drugs. The practice of searching for valuable natural resources is called **bioprospecting**. Extinction of species reduces the potential for important bioprospecting outcomes.

Various conservation strategies are in place to help maintain ecosystems and species of organisms, reduce damage and restore damaged ecosystems.

- **Conservation triage** aims to determine the order in which help is applied to vulnerable species based on their value. Value may rest on a species conservation status (for example, near extinction), ecological importance, or their cultural, social or economic importance to humans. Because financial resources are limited for conservation measures, this approach may mean some species are abandoned to extinction.
- **In situ conservation** refers to conservation strategies applied to species in their natural environment, such as preserving the habitat, eliminating invasive species or restoring damaged ecosystems.
- **Ex situ conservation** refers to strategies to conserve species outside of their natural habitat, for example, captive breeding programs in animal sanctuaries and zoos, or conserving genetic variation by establishing collections of seeds and other plant tissues.

Other strategies include identifying areas marked as conservation **reserves**.

- Reserves are areas set aside under government protection.
- In the protected environment of reserves, habitats and communities, including the species they support, can be monitored.
- Reserves don't have to be enormous; some species survive extremely well in remnant strips of vegetation along roads.
- Vegetation strips also act as corridors or links to larger reserves.

Restoration of disturbed or degraded ecosystems is another important approach aimed at rehabilitating habitats for species. Examples of degraded ecosystems include mining sites, logged forests, agricultural land and polluted waterways.

WORKSHEET 4.1

Knowledge review—essence of ecology

This activity introduces some of the key terms and concepts in this module by revisiting relevant foundation ideas you have considered before. You will recognise some of these from Module 3: Biological diversity, which have paved the way for the themes considered in Module 4: Ecosystem dynamics.

The words and parts of words in science often give us clues about their meanings. For example:

keystone

eco: environment, surroundings

bio: living

keystone: the central or keystone in a stone arch that gives integrity to the whole structure.

Apply your prior knowledge and understanding to write a logical definition of the following terms:

a ecosystem

b biodiversity

c keystone species

d biotic factor

e abiotic factor

f producer

g consumer

ISBN 978 1 4886 1931 1

WORKSHEET 4.2

Ecosystem epithets—naming ecosystems

Ecosystems are shaped in the first place by the abiotic factors to which the environment is subjected. Only organisms with adaptations suited to the particular conditions can survive there.

Four different kinds of ecosystems are represented in the following table. Think about the impact abiotic factors are likely to have on the development of each ecosystem and how this in turn influences the kind of community living there. Complete the table to discriminate between the different kinds of ecosystems, the abiotic factors that shape them and the kinds of organisms that inhabit them.

Ecosystem		Name	Key features	Abiotic factors that have helped to shape the ecosystem	Organisms in this community (list at least five)
a					
b					
c					
d					

RATING MY LEARNING	My understanding improved	Not confident ◄——► Very confident ○ ○ ○ ○ ○	I answered questions without help	Not confident ◄——► Very confident ○ ○ ○ ○ ○	I corrected my errors without help	Not confident ◄——► Very confident ○ ○ ○ ○ ○

WORKSHEET 4.3

Matchmaker—a web of interactions

1 Read the definitions listed in the boxes on the right of the page. Choose the correct term from the list below for each definition and write this term in the corresponding box.

parasitism	trophic level	consumer	commensalism	food chain	herbivore
scavenger	symbiosis	food web	mutualism	detritivore	keystone species
carnivore	decomposer	producer	amensalism		

Term	Definition
	a species that has a critical role in maintaining the integrity of an ecosystem
	linear feeding relationship between organisms in an ecosystem
	organism that uses photosynthesis to produce its own organic compounds
	organism that breaks down the dead remains of other organisms
	relationship in which one organism benefits at the expense of its host
	flesh-eating organism
	feeding level of an organism
	consumer that feeds on and breaks down decaying matter
	network of interconnected food chains
	organism that feeds only on plant matter
	organism that acquires its organic compounds by feeding on other organisms

2 Five terms remain. For each of these write a sentence summarising what is meant by the term.

Term 1: ______________________

Term 2: ______________________

Term 3: ______________________

Term 4: ______________________

Term 5: ______________________

RATING MY LEARNING	My understanding improved	Not confident ◄—► Very confident ○ ○ ○ ○ ○	I answered questions without help	Not confident ◄—► Very confident ○ ○ ○ ○ ○	I corrected my errors without help	Not confident ◄—► Very confident ○ ○ ○ ○ ○

ISBN 978 1 4886 1931 1

WORKSHEET 4.4

Wild web—living in an ecosystem

Ecosystems are characterised by the integration of various biotic and abiotic factors. Use the spaces provided to:

- name each organism
- describe its habitat in the ecosystem
- outline its ecological niche.

RATING MY LEARNING	My understanding improved	Not confident ◄──► Very confident ○ ○ ○ ○ ○	I answered questions without help	Not confident ◄──► Very confident ○ ○ ○ ○ ○	I corrected my errors without help	Not confident ◄──► Very confident ○ ○ ○ ○ ○

WORKSHEET 4.5

Fluctuating figures—population dynamics

The European carp (*Cyprinus carpio*) is an introduced species of fish that is threatening waterway ecosystems in Australia. The carp breed prolifically and outcompete native species for resources. The following table represents data for a lake that became infested with European carp, probably from eggs brought in on the hull of a fishing boat. The lake was previously free of the species despite being fed by a small tributary. Early data suggests that about five fingerlings (young fish) formed the initial population.

Year	Estimated population
1	500
2	3500
3	7000
4	15000
5	25000
6	32000
7	35000
8	38000
9	37000
10	39000

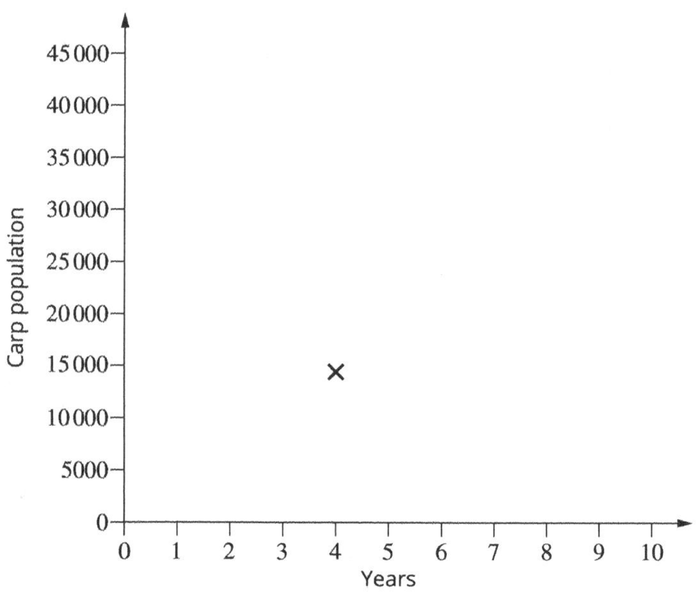

1 Use the tabulated data to graph the population growth of the carp over 10 years.

2 Name the kind of population growth that is occurring at point X on the graph.

3 Outline the factors that may contribute to growth in this population.

4 **a** Identify the part of the graph that represents the carrying capacity of this ecosystem.

b Explain what is meant by the carrying capacity of the ecosystem.

c Outline the factors that eventually limit population growth.

RATING MY LEARNING	My understanding improved	Not confident ○ ○ ○ ○ ○ Very confident	I answered questions without help	Not confident ○ ○ ○ ○ ○ Very confident	I corrected my errors without help	Not confident ○ ○ ○ ○ ○ Very confident

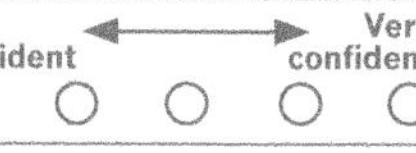

ISBN 978 1 4886 1931 1

WORKSHEET 4.6

Evolutionary ecosystems—past change

Ecosystems are dynamic in nature, with change occurring along a continuum. Small-scale changes that occur over very short periods of time tend to be less obvious than large-scale change that has occurred over long periods. Close examination of ecosystems reveals a rich story of change. The task of the scientist is to know where to look and what questions to investigate.

Examine the images below, then answer the questions about what they reveal about past change in the ecosystems they represent.

1 This Aboriginal rock art discovered at Ubirr in Kakadu National Park in the Northern Territory has been radiometrically dated to between 1500 and 28000 years old.

a Suggest the kind of animal this rock art depicts. Give reasons for your answer.

b What does this ancient art suggest about change in the distribution of this species over time?

c How might we account for this change?

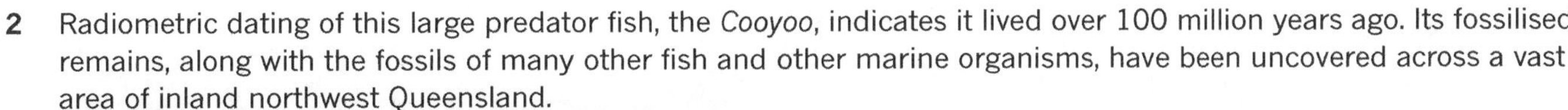

2 Radiometric dating of this large predator fish, the *Cooyoo*, indicates it lived over 100 million years ago. Its fossilised remains, along with the fossils of many other fish and other marine organisms, have been uncovered across a vast area of inland northwest Queensland.

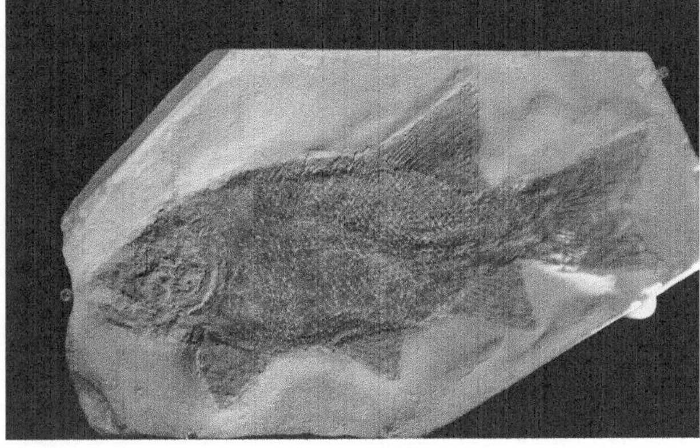

a What does this evidence suggest about ecosystem change in this region over the last 100 million years?

b Describe a change in an abiotic factor that may have led to ecosystem change and extinction for this species.

3 Analysis of air bubbles trapped in Antarctic ice and extracted in ice core samples is represented in the graph.

Change in temperature (°C)
0.8
0.6
0.4
0.2
0.0
Temperature
CO_2
370
290
CO_2 level (parts per million)
1850 1900 1950 2000
Year

a Describe the changes in atmospheric temperature between the years 1850 and 2000.

b Describe the changes in atmospheric carbon dioxide concentration in this time.

c Outline the relationship the graph suggests between atmospheric carbon dioxide concentration and temperature.

4 The photograph below illustrates coral bleaching of an area of the Great Barrier Reef. When coral polyps are stressed, the algae they contain are expelled and the coral becomes white or bleached. Without the algae, the life of the coral is threatened. A significant decline in corals has been measured between 1985 and 2010. Scientists believe this is largely due to cyclones, ocean acidification, warmer waters and the crown-of-thorns sea star.

Suggest how rising atmospheric carbon dioxide concentrations contribute to the decline of the coral reef.

RATING MY LEARNING	My understanding improved	Not confident ◄—► Very confident ○ ○ ○ ○ ○	I answered questions without help	Not confident ◄—► Very confident ○ ○ ○ ○ ○	I corrected my errors without help	Not confident ◄—► Very confident ○ ○ ○ ○ ○

 ISBN 978 1 4886 1931 1

WORKSHEET 4.7

Diminishing diversity—saving species

The impact of human activity on the natural resources and biodiversity of planet Earth has been devastating over the last 240 years. Growing human populations, together with land clearing, farming practices, urbanisation and industrialisation, have led to alarming statistics in relation to habitat and species loss on our planet.

In Australia alone:

- 30 species of mammals and 24 species of birds have become extinct
- more than 400 species of animals are currently listed as threatened
- more than 1200 plant species are currently listed as threatened.

Investigate an Australian species of plant or animal that is listed as threatened. Visit an Australian or New South Wales department of environment website to research the following questions. Be prepared to go to more than one site if necessary. Ask your teacher for assistance if needed.

1 Identify the species you have selected. ______________________________

2 Record the conservation status of this species and explain why it has this status. (Choose from Vulnerable, Endangered, Critically Endangered, and Extinct in the Wild.)

____

____

3 Identify and describe the human activities that have posed a threat to this species.

____

____

4 Describe the measures that are currently in place to address the threats to the species you have chosen.

____

____

5 Outline why this species is important.

____

____

____

6 Explain why conservation of species and ecosystems is an important issue for us to address.

____

____

____

____

RATING MY LEARNING	My understanding improved	Not confident ◄──► Very confident ○ ○ ○ ○ ○	I answered questions without help	Not confident ◄──► Very confident ○ ○ ○ ○ ○	I corrected my errors without help	Not confident ◄──► Very confident ○ ○ ○ ○ ○

ISBN 978 1 4886 1931 1

WORKSHEET 4.8

Literacy review—concise communication

In this module you have considered three central themes, which are listed below, along with some of their key terms. Communication of ideas is most effective when the relevant terminology is paired with considered sentence construction.

1 Select from the terms listed to complete the summary of population dynamics.

consumer	quadrat	intraspecific	community	abundance
ecosystem	decomposer	mutualism	carrying capacity	commensalism
transect	food web	parasitism	niche	producer
amensalism	abiotic	food chain	interspecific	

An ______________ is a system of organisms interacting with one another and with their non-living surroundings. Ecosystems are shaped by the ______________ factors that exist in particular regions, for example, topography and climate. An ecosystem supports a ______________ —a group of different species living and interacting together. Interactions between organisms within a species are referred to as ______________. Interactions between organisms of different species are ______________. There is a wide range of different kinds of interactions between organisms. Some relationships, such as ______________, are beneficial to both organisms. Some are beneficial to one organism but harmful to the other, for example, ______________ and ______________. ______________ is beneficial to one while the other remains unaffected. Every organism in an ecosystem has a role—this is called its ______________. An organism's role in the ecosystem contributes to the stability of the ecosystem, for example, ______________ organisms such as plants manufacture organic compounds during photosynthesis. ______________ break down organic materials in the dead remains of organisms, recycling them to be used again. ______________ organisms rely on producers for these organic compounds. A linear feeding relationship is called a ______________ ______________. A network of food chains forms a ______________ ______________. Several factors contribute to the density and ______________ of organisms in ecosystems, for example, availability of resources such as food and shelter. However, ecosystems have a limit to the number of organisms they can support—this maximum population size is called its ______________. Sampling techniques are used to monitor populations in ecosystems. A ______________ provides data about population density in marked areas, while a ______________ provides information about the distribution of organisms along a line marked through the area. Populations in ecosystems fluctuate in response to many factors. When these factors are out of balance, the implications can be significant.

2 The following summary of past ecosystems is incomplete. Select from the terms listed to complete the statements, then incorporate the remaining terms in your own statements to complete the summary. It will be helpful to use only one or two terms in each sentence. Underline or highlight these key tems as you include them in your summary.

disturbance	megafauna	biosecurity	fragmented habitat
biodiversity	invasive species	selection pressure	bleaching
extinction	Great Barrier Reef	Intermediate Disturbance Hypothesis	

Ecosystems are dynamic and constantly undergoing change. Stable ecosystems maintain their integrity when key factors are in balance. For example, populations of organisms remain stable due to equilibrium between rates of birth and death, food demand and food supply. Ecosystems maintain their stability through feedback loops in which populations fluctuate in response to factors such as available resources. A ______________ in an ecosystem is a factor that drives significant change. The impact of the change depends on the size of the disturbance. The ______________ ______________ ______________ proposes that high and low levels of disturbance decrease the diversity of ecosystems.

 ISBN 978 1 4886 1931 1

3 Prepare a carefully constructed summary of future ecosystems using all the terms listed below.

genetic variation
species distribution model
Representative Concentration Pathways (RCPs)
anthropogenic pressure
conservation
pollution
bioprospecting
bioresources
restoration
bioindicators

RATING MY LEARNING	My understanding improved	Not confident ◄──► Very confident ○ ○ ○ ○ ○	I answered questions without help	Not confident ◄──► Very confident ○ ○ ○ ○ ○	I corrected my errors without help	Not confident ◄──► Very confident ○ ○ ○ ○ ○

ISBN 978 1 4886 1931 1

WORKSHEET 4.9

Thinking about my learning

On completion of Module 4: Ecosystem dynamics, you should be able to describe, explain and apply the relevant scientific ideas. You should be able to work with data, to interpret, analyse and evaluate it.

1 The following table lists the key knowledge covered in this module. Read each and reflect on how well you understand each concept. Rate your learning by shading the circle that corresponds to your level of understanding for each concept. Remember to use colour as a visual representation if you find this helpful. For example:

- green—very confident
- orange—in the middle
- red—starting to develop.

Concept focus	**Rate my learning** Starting to develop ◄——► Very confident				
Relationships between biotic and abiotic factors in ecosystems and the impacts of these factors	○	○	○	○	○
Ecological niches occupied by species	○	○	○	○	○
Consequences of selection pressures such as predation, competition and disease for populations	○	○	○	○	○
Extinction and extinction events	○	○	○	○	○
Palaeontological and geological evidence for past changes in ecosystems	○	○	○	○	○
Technologies used to analyse change in ecosystems	○	○	○	○	○
Causes of short-term and long-term change in ecosystems	○	○	○	○	○
Human impact on ecosystems, for example, climate change and extinctions	○	○	○	○	○
Conservation management strategies	○	○	○	○	○

2 Consider points you have shaded from starting to develop to middle-level understanding. List specific ideas you can identify that were challenging.

____

____

____

3 Write down two different strategies that you will apply to help further your understanding of these ideas.

____

____

____

ISBN 978 1 4886 1931 1

PRACTICAL ACTIVITY 4.1

Plants in their place—using line transects

Suggested duration: 50 minutes

INTRODUCTION

When we look around us, we may see a diversity of plants growing in a community. As biologists, it is useful to describe plant communities. This can be done in a variety of ways. We will often give a community a name based on the most obvious or common plant species present (the dominant species), for example a mountain ash forest. We can make a more detailed description by listing all the plant species present and estimating their abundance. This activity and the next look at two different ways of making this estimation. They are:

- a detailed estimation by using a line transect (this activity)
- an estimation made by looking at the community and making a rough count of the numbers of each species present (Practical activity 4.2).

MATERIALS

- clipboard
- paper and pencil
- measuring tape (at least 10 m long) or a line with distances marked on it
- metre ruler
- labels for plant specimens—masking tape is suitable
- secateurs (optional)
- reference material to assist with identification (optional)
- camera (optional)

PURPOSE

To use two different methods to describe the composition of a plant community.

To make some comparisons between the results obtained by using the different methods.

BACKGROUND

In the course of their work, many people seek to describe the vegetation patterns of a particular area.

- An ecologist studies the recovery of an area of bush burnt in a bushfire.
- A wildlife ranger investigates changes in vegetation as a result of increasing kangaroo numbers in Bouddi National Park.
- A biologist describes an area of tropical rainforest for the first time.
- A farmer wishes to re-establish native vegetation on some cleared marginal land.

The reason for studying vegetation in this way is usually to find answers to one or more of three questions:

- What plant species are present in the area?
- How abundant are the different plant species?
- How are the different plant species distributed?

This background provides vital information about the community under investigation.

A walk through an area being studied would probably reveal many of the larger plant species, but a detailed investigation will be needed to answer questions about abundance and distribution of all the plant species.

Biologists use several techniques to obtain data from which they can answer these questions.

Transects

Transects are lines or belts set out through an area as a guide for recording what plant species are present. They provide a useful method for assessing changes in the abundance of a particular species in response to changes in a physical variable in the environment, such as slope gradient or soil moisture content, as shown in the following figure.

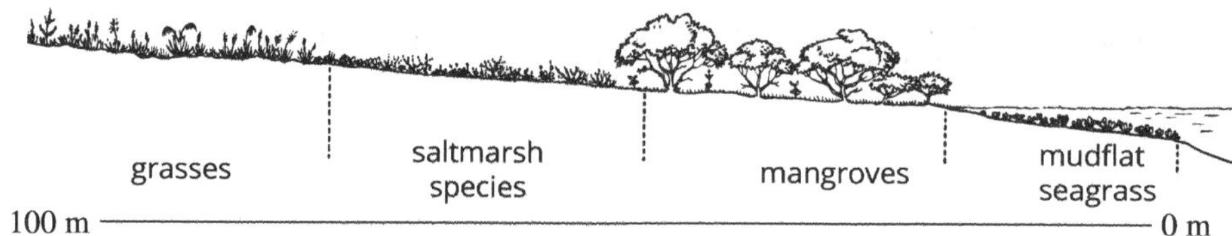

Transect line showing location of seagrass, mangroves and saltmarsh plants and grasses

Line transect

A line transect is made by running out a tape measure or marked line across the area to be sampled. The observer works systematically along the line, recording the name and position of each plant that the line passes over, under or through. The data is recorded on paper on a corresponding line drawn to scale.

Belt transect

This is similar to a line transect except that all the vegetation between two parallel lines is recorded. In a way it is like a long, thin quadrat. A transect can also be used as a location line for quadrats, in which the cover of each species can be estimated. (Quadrats are discussed further in Practical activity 4.2.)

Profile diagram

A profile diagram is a scale drawing of the profile, or 'side view' (also called the elevation), of the vegetation along a line. It may also show the shape of the land surface and details of soil type. Care needs to be taken with the choice of the vertical scale so that the drawing is not too distorted. Profile diagrams are able to show the horizontal layers or strata of the vegetation. Well-drawn profiles are often useful for making qualitative comparisons between different plant communities.

(a)

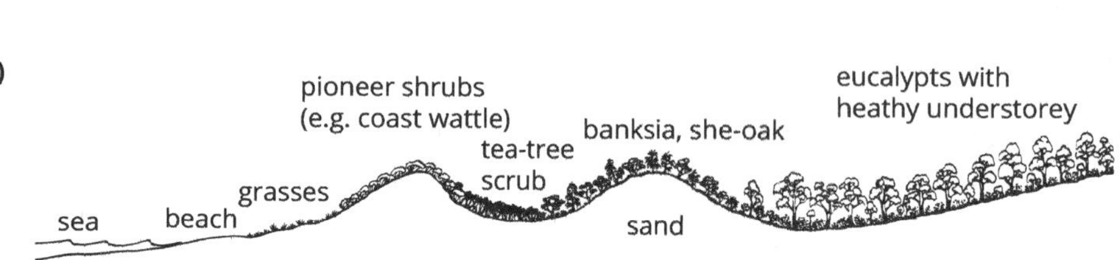

(a) Profile diagram of vegetation growing on sand dunes

(b)

(b) Detailed profile diagram of a section of a wet sclerophyll forest

Identification of plants

When studying a plant community it is important to have a ready means of identifying the species present. A commonly used method is to make use of a field herbarium. A field herbarium contains specimens of each of the plants in the area, collected and pressed between sheets of paper. The important parts to collect are those that are useful for identification in the field, such as leaves, flowers and fruits. When dry, these are mounted on sheets of card, labelled and stored in plastic sleeves. Plants that cannot be readily identified can be labelled as species A, species B, and so on. Good quality colour photographs are also useful aids for identification in the field.

ISBN 978 1 4886 1931 1

PROCEDURE

1 Walk through the area you intend to describe and collect samples of each plant species present. Make sure that you are permitted to collect in the area you have chosen. Take care when collecting to see that you do not cause unnecessary damage, and use secateurs if they are available. You only need a small piece of each species.

2 Put each plant species into a category that describes its life form. You do not need to identify the species by scientific name. You might find the following categories useful:

- Trees: ecologists define trees as woody plants over 5 m tall, usually with a single stem. Label as 'tree 1', 'tree 2', etc.
- Shrubs: ecologists define shrubs as woody plants less than 5 m tall, frequently with many stems arising at or near the base. Label as 'shrub 1', 'shrub 2', etc.
- Herbs: ecologists define herbs as any non-woody plant. Because the group could be very large, further subdivision into ferns, grasses and broad-leaf plants may be helpful. Label as 'fern 1', 'fern 2', 'grass 1', 'grass 2', 'broad-leaf 1', 'broad-leaf 2', etc.
- Climbers: label as 'climber 1', 'climber 2', etc.

Use whichever categories you find easiest to work with, but be consistent with the categories in which you place the different plants. Label your specimens using your chosen system so that you can refer back to them later.

3 Each species is to be given a ranking number, in order of abundance. The most abundant (common) species is ranked 1. The least abundant (rarest) species is ranked with the highest number.

By group decision, rank each plant from most common to rarest.

PROCESSING DATA

1 List each plant you have labelled in the 'Species' column of Table 1.

2 Enter the rankings next to the plant names under the column headed 'First ranking'.

TABLE 1 Ranking species

Species	First ranking	Line transect count	Second ranking
tree 1			
shrub 1			

PROCEDURE

4 Now run out the tape measure or line in a straight line through the area. Move along the tape from one end and, for each plant species, count the number of individual plants of that species that the tape passes over, under or through. The easiest way to do this is to select a particular species and walk along the line counting the number of that species. Then select the next species and repeat the procedure. Continue to do this until all species have been counted.

PROCESSING DATA

3 Record your counts for each species listed, in the 'Line transect count' column of Table 1. You have now made a line transect through the study area. It may be difficult to count the individual grass plants along the transect.

4 Describe another method of comparing the abundance of grasses with that of other species along the line.

PROCEDURE

5 If time permits, make a second line transect through the same area.

6 Using the same system of ranking as used in step 3, make up a new ranking order based on the data in the 'Line transect count' column of Table 1. Some plant species may not show up in your count; ignore these for the moment.

PROCESSING DATA

5 Record these rankings in the 'Second ranking' column of Table 1.

CONCLUSION

6 a Describe any differences you noted between your two rankings.

b Suggest possible reasons for the differences.

7 a List any species that were present in one ranking and absent from the other.

b If any species were absent in one of the rankings, suggest why this might be.

 ISBN 978 1 4886 1931 1

8 Suggest why it is better to collect data from more than one transect.

9 Which method best describes the area you have studied? Explain your reasoning and draw a diagram to support your answer.

RATING MY LEARNING	My understanding improved	Not confident ⟷ Very confident ○ ○ ○ ○ ○	I answered questions without help	Not confident ⟷ Very confident ○ ○ ○ ○ ○	I corrected my errors without help	Not confident ⟷ Very confident ○ ○ ○ ○ ○

PRACTICAL ACTIVITY 4.2

The flatweed census—population estimation

Suggested duration: 100 minutes; about 50 minutes in the field and about 50 minutes to complete calculations and written report

INTRODUCTION

Your lawn is covered in weeds. Will regularly digging out the dandelions really reduce their abundance? Does spraying the lawn with weed killer reduce the number of weeds more effectively than digging them out? What experimental techniques can you use to answer these questions? How can you accurately record the abundance of weeds in your lawn before and after digging or herbicide treatment?

MATERIALS

- quadrat (1 m^2 is a suitable size)
- tape measure (10 m)
- clipboard and paper
- weed species chart from your local council or from the Internet

PURPOSE

To use quadrats to estimate the population density of a flatweed species in a lawn.

To use these data to estimate the total population of the flatweed in the lawn.

BACKGROUND

Quadrats

A quadrat is a square, rectangular or circular frame of convenient size, used to mark out an area in which the vegetation is to be sampled. The shape and size of the quadrat depends on the type of vegetation. A square with 50 cm sides would be suitable for sampling a lawn, while a square with 10 m sides may be used for sampling trees in a forest.

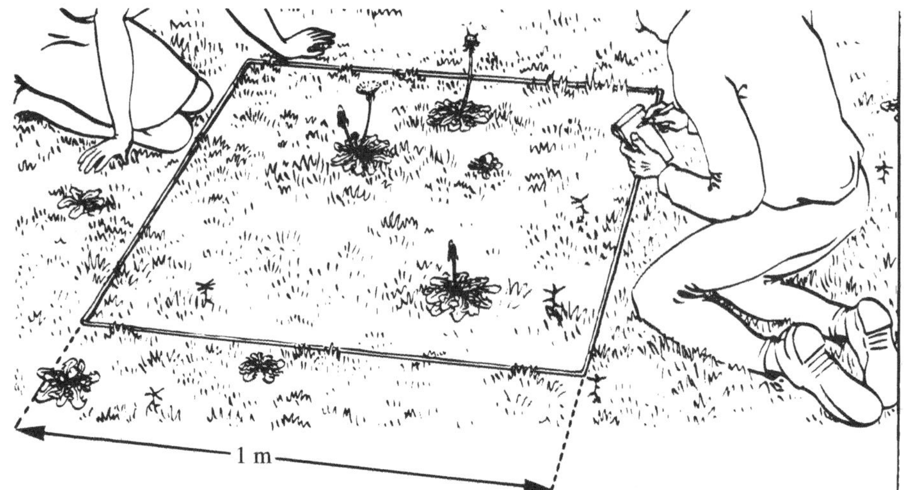

A 1 m^2 quadrat in use for sampling an area of lawn. In this case the density of dandelions is 4 per m^2 and the cover is 10 per cent of the area sampled.

When a quadrat is used to sample vegetation, first a list is compiled of all the plants contained in the quadrat. Repeating this for several quadrats should provide a composite list of the species present in the area. Quadrats can also be used to give an estimate of the abundance of one or more of the species.

Abundance is measured in two ways:

- Count the number of a particular plant in the quadrat. This gives the density of the plant, for example the number of dandelions per square metre.
- Cover is the percentage area of the quadrat covered by a particular plant species, for example the area of the quadrat that is covered by dandelion plants.

ISBN 978 1 4886 1931 1

PRACTICAL ACTIVITY 4.2

Quadrats may be located either randomly over the area being sampled, or at regular intervals along a transect or on a grid (see the following figure).

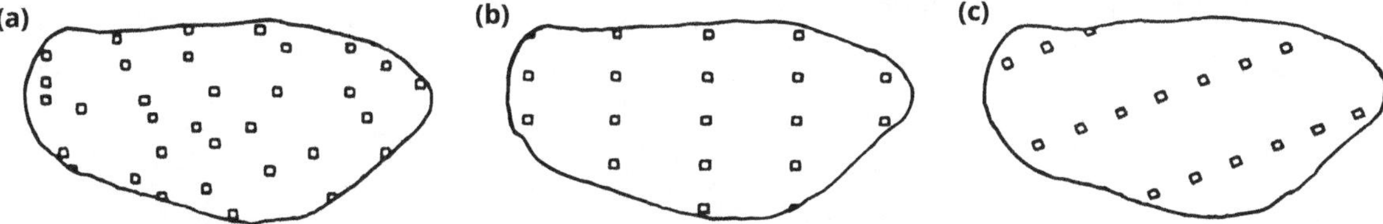

Methods of distributing quadrats: (a) random distribution, (b) spaced as widely and evenly as possible (grid), and (c) distributed evenly along transect lines

PROCEDURE

1 In groups, select a suitable area of 'lawn' that has populations of different flatweed species. Work out a way of determining the total area (in square metres) of the lawn you are studying.

2 Choose the flatweed that you want to count and, using a reference, try to identify it.

3 Discuss the following questions with others in your group:

  a What percentage of the total area would you need to sample in order to cover the lawn adequately?

  b How many quadrats should be used to do this?

4 Discuss with others in your group how you should locate the quadrats.

PROCESSING DATA

1 Prepare a sketch of the lawn selected for sampling. The sketch should show the general shape of the sample area and include measurements and an appropriate scale.

ISBN 978 1 4886 1931 1

2 Keep a record of the species you have chosen. Give each species a name, describe its features and draw a sketch to illustrate your description. Several groups should count the same species and then combine and average the results to improve the reliability of your data.

3 Record the number of quadrats the group agreed upon and explain how you arrived at this number.

4 Describe the method your group has chosen to locate the quadrats and explain why you have chosen this method.

PROCEDURE

5 Locate the first quadrat using the method outlined in your answer to step 4. Count the number of flatweed plants in the quadrat. For plants that the quadrat frame touches, use the following method: if more than half the leaf area of a plant is inside the quadrat, then count that plant; if less than half the plant is inside, then do not count that plant.

6 Repeat step 5 for at least five quadrats.

PROCESSING DATA

5 Record the count in Table 1.

TABLE 1 Separate group results

Quadrat number	Number of weeds
1	
2	
3	
4	
5	
Total	

6 A quadrat with four 1-m sides has an area of 1 m^2. If you used a 1 m^2 quadrat, the average number of flatweed plants per quadrat is the same as the average density, that is, the number of flatweed plants per m^2.

A quadrat with four 0.5-m sides has an area of 0.25 m^2. If you used a 0.25 m^2 quadrat then multiply your average number of flatweed plants per quadrat by 4 to calculate the average density.

Use your data from Table 1 to calculate the average number of weeds per quadrat. Show your working.

 ISBN 978 1 4886 1931 1

7 Using your measurement of the area of the lawn, calculate and record your estimate of the size of the flatweed population in the lawn. Show your working.

8 Record the data of other groups who counted the same flatweed species in Table 2.

TABLE 2 Combined results for groups using the same flatweed species

	Number of weeds counted (total of all quadrats)	Number of quadrats sampled	Average number of weeds per quadrat	Estimated flatweed population in the lawn
Group 1				
Group 2				
Group 3				
Group 4				
Total				

CONCLUSION

9 **a** Describe any differences in the estimates of the population size calculated using your group's data and the combined data.

b Suggest reasons for any differences.

c Which of the two estimates of population size do you expect is more accurate? Explain the reasons for your answer.

10 Outline how quadrats are useful in determining population density.

11 Evaluate the technique you used.

a Describe any limitations you encountered.

b Suggest how they could be overcome.

RATING MY LEARNING	My understanding improved	Not confident ◄——► Very confident ○ ○ ○ ○ ○	I answered questions without help	Not confident ◄——► Very confident ○ ○ ○ ○ ○	I corrected my errors without help	Not confident ◄——► Very confident ○ ○ ○ ○ ○

PRACTICAL ACTIVITY 4.3

In the field—a report on an ecosystem

Suggested duration: 100 minutes

INTRODUCTION

This activity provides an opportunity to conduct a fieldwork exercise in a selected ecosystem. A local terrestrial or aquatic ecosystem would be desirable and convenient; however, a field study of an ecosystem in the school grounds or even the classroom may be satisfactory. If you are able to visit a local nature reserve, remember that these are set aside by local authorities to protect native plant and animal species. Be mindful to create the least possible disturbance. Ensure that you have permission from the appropriate authority if you must collect specimens for analysis back in the school laboratory. Monitoring and analysis that can be completed in the field should be done there.

Field guides and keys for the identification of plants and animals will be useful to take along. Take care to protect yourself from stinging or biting insects and spiders when observing particular parts of the ecosystem. For example, use a stick or pencil to gently lift the bark of trees or to lift leaf litter. Rubber gloves may be useful. Ensure rocks that have been upturned during the investigation are returned to their original position.

You need to be familiar with different kinds of aquatic and terrestrial ecosystems, so background reading is recommended. In this activity you will classify the ecosystem you study. You also need to be able to choose and conduct appropriate sampling techniques for estimating population sizes for plant and animal species. Practical activities 4.1 and 4.2 will be useful background activities.

You are likely to be working as part of a team, contributing part of the overall data to the investigation of the ecosystem, and relying on others to provide data as well. Therefore, it is important to keep an accurate record of all of the data you collect.

MATERIALS

- pencils
- measuring tape (at least 10 m long)
- rope or cord
- 1 m × 1 m (1 m^2) quadrat
- thermometer or temperature probe
- light meter
- anemometer (or devise a suitable scale, e.g. 5 point)
- pH meter or probe
- secchi disk or turbidity tube
- oxygen meter or probe
- magnifying glass
- field guides, e.g. of grasses, herbs, eucalypts, invertebrates
- extra paper for notes and diagrams, if required

PURPOSE

To complete a fieldwork exercise in a selected ecosystem.

To consider the biotic and abiotic factors that make up an ecosystem.

To estimate the abundance and distribution of selected organisms within an ecosystem and consider the factors that might affect them.

To consider some of the relationships between species within an ecosystem.

PROCEDURE

In the field

In the field you will be actively collecting data. This will include making notes and diagrams.

1 When you arrive at the chosen study site, take a few minutes to look around. Observe the main features, prominent landmarks, the kinds of vegetation in the area, dominant plant species and weather conditions.

2 Observe the ecosystem closely, looking out for any plant and animal life.

3 Stand or sit quietly (for at least five minutes) to determine the kinds of animals in the ecosystem. Use other senses as well as your sight—listen for sounds. Examine the bark of trees where tiny animals may be hidden or camouflaged. Look for evidence of the presence of animals—for example nests, spider webs, scats, footprints, trails, burrows or chewed leaves.

ISBN 978 1 4886 1931 1

PRACTICAL ACTIVITY 4.3

PROCESSING DATA

1 Write a detailed description of the ecosystem, including abiotic and biotic features. Abiotic features will include geographical characteristics as well as local weather conditions. When describing biotic features, mention the type and density of vegetation. Consider features that will help you identify the kind of ecosystem you are in. For example, for terrestrial ecosystems consider the density of grass and/or trees (grassland or forest) and the amount of sunlight that penetrates the foliage of trees to reach the ground. For aquatic ecosystems think about if it is a saltwater, freshwater or estuarine ecosystem. Is the body of water a lake, stream, river or sea? Give a name to the kind of ecosystem you are studying.

2 Make two diagrams on this and the following page to represent the area—a plan sketch and a profile sketch. Either label your diagrams to indicate various features or use a key. Use the following figures as a guide.

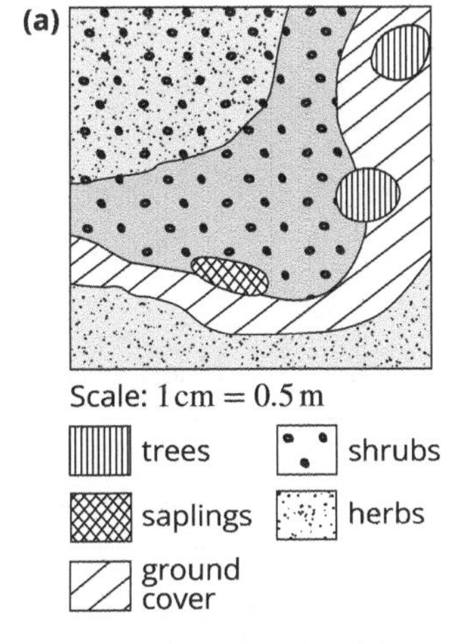

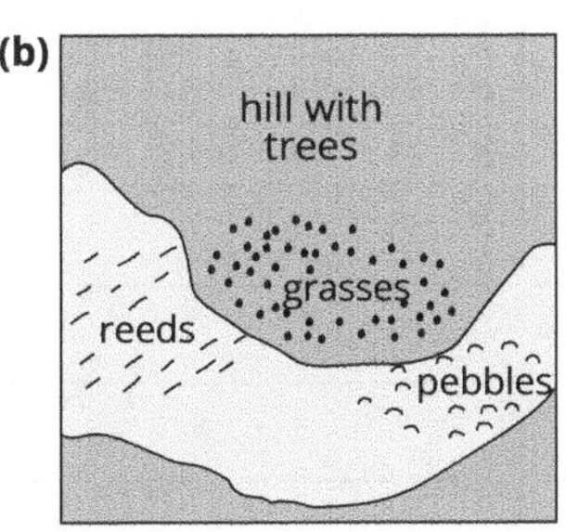

Plan sketch of (a) a terrestrial and (b) an aquatic ecosystem

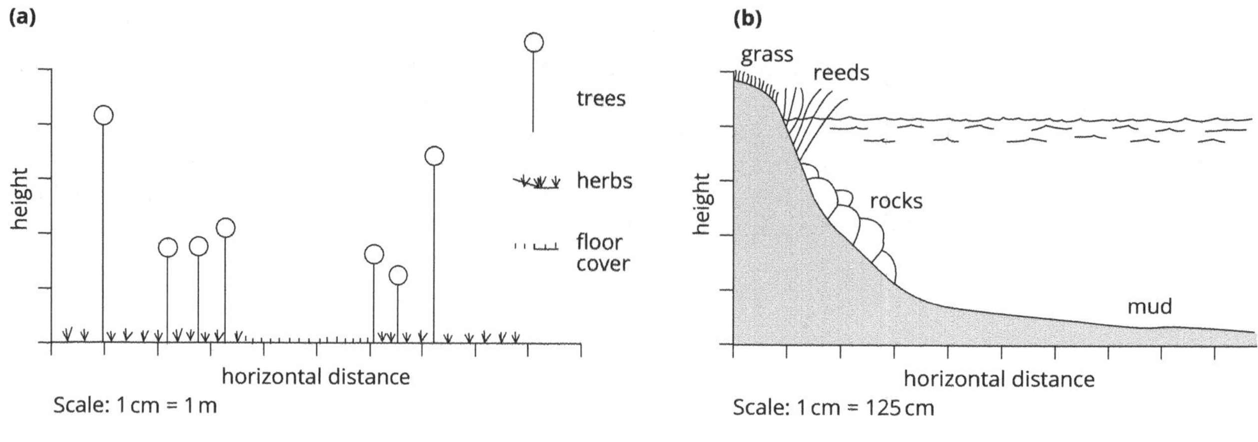

Profile sketch of (a) a terrestrial and (b) an aquatic ecosystem

Abiotic factors

3 Measure as many abiotic factors as possible for your ecosystem. Record the results of each test in Table 1 as you go.

TABLE 1 Summary of abiotic factors

Abiotic factor	Measurement
air temperature • at ground level • at 1 m above ground level	
soil temperature	
water temperature	
light intensity	
wind speed	
soil pH	
water pH	
water turbidity	
dissolved oxygen content of water	
flow rate of water (place an object in the water and measure distance covered over a given time, e.g. metres per minute)	

 ISBN 978 1 4886 1931 1

4 Identify and describe any other abiotic factors in the ecosystem.

Biotic factors

5 Write a list of the different kinds of vegetation in the ecosystem, for example, trees, grasses, herbs, ferns, mosses and lichens. Also include algae and fungi that may be present. Provide specific names of plants that you are able to identify.

6 Prepare a list of the types of animals evident in the ecosystem. Identify as many as you can.

7 Use your knowledge of interactions between organisms within ecosystems and your observations to draw four different food chains for this ecosystem.

8 Draw a food web for this ecosystem that includes at least two different plants and at least four different animal species. You could start by combining the food chains you constructed in question 7.

PROCEDURE

Population abundance and distribution

In this part of the fieldwork you will use some sampling techniques to gather data related to the distribution and abundance of organisms.

4 Use the rope and measuring tape to mark a transect line at least 10 m long through a selected area of the ecosystem. Make sure your transect line runs through an area that varies over the distance selected. For example, the vegetation may change, there may be a body of water or undulations in the landscape.

5 Walk the length of the transect line, observing the plants and animals along the way.

PROCESSING DATA

9 Draw a profile sketch of the ecosystem represented by the transect line (see the figures on page 180 for examples of profile sketches).

10 Complete Table 2, entering the different plant and animal species you observe along the transect line.

TABLE 2 Plant and animal species observed along transect line

Plant species	Animal species

ISBN 978 1 4886 1931 1

PROCEDURE

6 Choose a plant species from your list. Set up a quadrat 1 m × 1 m at one end of the transect line. Count the individual plants of your chosen species within the quadrat.

7 Repeat the procedure for another four quadrats set at intervals along the transect line to give a random selection of the plant type you are investigating. Enter the data into Table 3.

TABLE 3 Number of plants

Quadrat number	Number of plants

8 Discuss with other members of your team how to estimate the abundance of this plant species in the area of the ecosystem you are studying. Decide where to locate another five quadrats in order to achieve meaningful results. Count the plants and add the data to Table 3.

ANALYSIS OF RESULTS

11 a Describe the distribution of the plant species along the transect line.

b Account for any differences in the distribution you have observed.

12 Use the data in Table 3 to calculate the average size of the plant population per quadrat. Show your working.

13 Estimate the population size (number of individuals) of the plant species in the area. Show your working.

14 Comment on the variability of the measurements made in different quadrats.

PROCEDURE

9 Choose a small animal, for example an insect species, from the list of animals you compiled in question 6. With the members of your team discuss a suitable way of estimating the abundance of this species in the area.

ANALYSIS OF RESULTS

15 a Describe an obvious problem and its consequences when monitoring and recording animal numbers (especially very small ones, such as ants) in a sample area.

b Outline a more effective approach to monitoring animal numbers.

16 Estimate the abundance of the animal species in the area. Indicate how you arrived at the estimation. Record the date and time you made the estimate, along with a description of particular circumstances that might affect the abundance of the animals at the time.

17 Describe the distribution pattern of this animal species in the area. (Hint: look at the features of the ecosystem that provide the optimum conditions for the species—for example shade plants, rocks for shelter, availability of food.)

PROCEDURE

Human impact

10 Look carefully around the ecosystem for signs of human impact. This may include deliberate change or indirect effects resulting from human activity outside the ecosystem. Examples of human activity could include pollution (for example air, water and noise), clearing of land, building, fencing, recreational use of land/water, or introduced species of plants and animals.

ANALYSIS OF RESULTS

18 Describe how human activity appears to have impacted on the ecosystem.

 ISBN 978 1 4886 1931 1

PRACTICAL ACTIVITY 4.3

19 a Describe any limitations you encountered in undertaking this fieldwork.

b Suggest how they could be overcome for subsequent fieldwork.

SUMMARY

20 a Suggest how the abiotic factors in this ecosystem are likely to change over a 24-hour period.

b Describe how such changes in the abiotic factors are likely to affect the plant and animal life in the ecosystem.

21 Describe an example of each of the following kinds of relationships evident in the ecosystem:

a predator—prey

b parasite—host

c mutual benefit

PRACTICAL ACTIVITY 4.3

22 Use your knowledge of food webs to explain how energy moves through ecosystems.

23 Outline some factors that might account for the distribution and abundance of organisms in the ecosystem.

24 Draw a concept map summarising the relationships between organisms within the ecosystem and between the organisms and their non-living surroundings. Include inputs and outputs relevant to the ecosystem.

ISBN 978 1 4886 1931 1

PRACTICAL ACTIVITY 4.3

25 Considering the human impact on this ecosystem discussed earlier in this activity, suggest strategies that might lead to an improvement in the quality of this ecosystem.

26 Discuss the importance of protecting and maintaining ecosystems.

RATING MY LEARNING	My understanding improved	Not confident ◄——► Very confident ○ ○ ○ ○ ○	I answered questions without help	Not confident ◄——► Very confident ○ ○ ○ ○ ○	I corrected my errors without help	Not confident ◄——► Very confident ○ ○ ○ ○ ○

DEPTH STUDY 4.1

Cascading communities—the ripple effect of change in ecosystems

Suggested duration: Part A—50 minutes; Part B—3 hours

INTRODUCTION

Ecosystems are characterised by complex interactions between organisms, and between organisms and their non-living surroundings. Ecosystems tend to remain relatively stable when low-level disturbances operate. High-level disturbances can have profound effects in ecosystems, causing rapid and dramatic change. Yellowstone National Park in the United States represents a remarkable example of profound change in an ecosystem over a relatively short time.

In 1995, 14 wolves were reintroduced into the Yellowstone National Park ecosystem after an absence of 70 years. It wasn't until their reintroduction and the trophic cascade that followed, that ecologists understood more deeply the critical role this higher order predator played in the stability and maintenance of the wider ecosystem. A trophic cascade is an ecological process that occurs when a predator's presence or absence alters the behaviour of its prey, creating a ripple effect that alters interactions between biotic and abiotic factors, ultimately causing far-reaching ecological change.

Ecologists continue to learn about such change from the events still unfolding in Yellowstone National Park. Observations of such events raise many questions about the fundamental nature of ecosystem dynamics, and the potential impact that human decision-making and actions can have.

PURPOSE

To consider the impact one species can have on other species in a community.

To conduct your own investigation on changing dynamics in an ecosystem due to the influence of a single species.

To predict outcomes for species resulting from ecosystem change.

 ISBN 978 1 4886 1931 1

Part A—the wolves of Yellowstone National Park

CONDUCTING YOUR INVESTIGATION

Using the search terms 'reintroduction of wolves to Yellowstone National Park', find a short (four to five minutes) video that will help you learn more about the dynamics of this ecosystem.

PROCESSING DATA AND INFORMATION

1 Watch the video a couple of times. On the second viewing record summary notes related to:

- ecosystem
- biotic factors
- abiotic factors
- interactions between organisms
- how the absence of the wolves drove change in Yellowstone National Park.

ANALYSING DATA AND INFORMATION

2 a Draw a mind map to illustrate the cascade of changes that occurred in Yellowstone National Park as a result of the reintroduction of wolves.

b Highlight and describe any trends, patterns or relationships that you observe in your mind map.

 ISBN 978 1 4886 1931 1

3 Identify an abiotic factor altered by the presence of the wolves after their reintroduction in 1995. Describe how this factor changed.

QUESTIONING AND PREDICTING

4 The impact of wolves in the Yellowstone National Park ecosystem is far-reaching and observable in some rather unexpected ways. It raises many questions about the interactions between the organisms that live there, between those organisms and their non-living surroundings, and the magnitude of the effects a single species can have. Two questions about these interactions are set out below. Construct two more questions raised by the video. Write them down. Use your mind map notes in question 2 to summarise the answer to the question.

- How do wolves impact on animals with which they have no direct interaction?
- How could the presence of wolves affect the stability of riverbanks?

Question	Answer

5 Predict possible longer-term consequences in Yellowstone National Park had the wolves not been reintroduced. Describe at least two.

ISBN 978 1 4886 1931 1

Part B—the ripple effect of a selected species

In Part B you will investigate the effect one species can have on another. You will select an introduced species that has effected change in an Australian ecosystem for your own independent research. You will ask questions, make predictions, gather data, record and communicate your findings in a digital presentation.

This research task offers an opportunity to explore more deeply the kinds of interactions that occur between biotic and abiotic factors in ecosystems and the profound effects such interactions can yield. You will consider these interactions and changes in the context of human activity that can precipitate change in ecosystems. You will consider the issues and propose realistic solutions. The final part of this presentation includes a written summary evaluating the human impacts and responsibilities for the stewardship of ecosystems.

PLANNING YOUR INVESTIGATION

Some examples of species that have effected change for other species are listed below. You may select one of these or choose another in consultation with your teacher. In making your selection, it is important to identify your aims. What information do you want to understand? The answer to this question will drive your investigation.

- crown-of-thorns sea star
- cane toad
- prickly pear
- European carp
- fox
- camel
- feral cat
- feral pig
- cattle in the Alpine National Park

COMMUNICATING

Criteria

You are required to address each of the following criteria in your investigation:

a Provide the common and scientific name of the selected species.

b Identify its country of origin.

c Summarise historical information around its introduction into Australia. Include intended reasons for its introduction and approximate dates if known.

d Use a map to show the progressive range of the species in Australia over time. Include a brief written summary of the information.

e Browse through your resources to obtain relevant statistical information about the species and its impact, if available. Include this information and explain what it means.

f Outline the impact of this species on the Australian ecosystem, giving particular attention to its interactions with and effects on native species. Include impacts on both biotic and abiotic factors where relevant.

g Construct a mind map to summarise these interactions.

h Identify two key questions that need to be asked to more fully understand the ecosystem dynamics in question.

i Predict the potential outcomes for native species if the current scenario continues.

j Describe conservation strategies that have been implemented to manage the problem and comment on the success or otherwise of these.

k Suggest other possible solutions to the problems associated with this species.

 ISBN 978 1 4886 1931 1

CONCLUSION

Summary

A summary concludes your investigation, bringing together key findings and evaluating human actions and responsibilities for the stewardship of ecosystems. Your response to the following points is expected to be approximately one page in length.

a What have you learnt about ecosystem dynamics and the impact one species can have on others?

b Outline the significance of biodiversity in this ecosystem, giving reasons why biodiversity protection is important.

c Comment on the importance of human awareness and responsibility in maintaining biodiversity and healthy ecosystems.

d How do past changes, such as those that have led to decreases in biodiversity or extinction of species, inform our future management of resources?

Guidelines

i You will communicate your findings about your selected species and its impact using a digital format of your choice, as well as a one page written summary. It is important that you adhere to the following protocols:

- Write your own notes from your research. Writing these in your own words demonstrates your understanding of the information.
- Use your research notes to plan the content, order and construction of your presentation.
- Use appropriate terminology to demonstrate you are familiar with the language of the subject and the topic.
- Define new terms for your audience in your own words.
- Use footnotes to reference factual information, scientific studies and statistical data.
- Prepare a list of references according to expected guidelines.

ii It will be helpful to refer to the Toolkit to guide you through the elements of a secondary-sourced investigation. You will be required to locate relevant and reliable resources, organise your notes, analyse the information you uncover and organise and present it in an appropriate fashion.

iii As a guide to length for this task, a PowerPoint presentation should be no more than 14 slides, including the title slide and references. The one page written summary forms the final part of the task.

iv Keep your work to the point. You will be assessed on addressing each criterion logically and clearly, not on volume.

Scope of activity

The examples listed above represent a range of impacts on Australian ecosystems. They also focus on the influence of introduced species on ecosystems and the organisms they support. The wolves of Yellowstone National Park are one of many examples that demonstrate the consequences of species removal, as well as introduction. Species can have profound impacts on ecosystems in their absence as well as their presence. With your teacher's guidance, you may decide instead to focus on one of the following inquiry questions or another of your choice.

- What happens when top predators such as sharks are overharvested from their marine ecosystem?
- How has overfishing in the Galapagos Islands impacted on marine and terrestrial ecosystems?
- How has overexploitation of sea turtles (for eggs, meat, skin, and shells) affected the marine ecosystem of which they are a part?

Conservation organisation websites are good resources for exploring other examples.

If you decide to pursue one of these questions instead, you will still follow the criteria and guidelines for undertaking this Depth study. However a couple of points from the criterion will require adjustment:

a Summarise historical information around the exploitation of the selected species.

b Use a map to show the declining geographic range of the species over time and make relevant adjustments to geographic references.

You may be interested in planning and conducting a primary investigation in which you focus on questions about change in an ecosystem and then plan and conduct an experiment, ahead of gathering data and drawing conclusions. An accessible example is to monitor change in the garden ecosystem at home. Think about what happens when a new patio or garden path is laid down. Leave it unattended for three months and the changes are profound. An entire ecosystem of plants and animals will be living there.

- Prepare questions about the ecosystem, for example:
  - How likely is it that organisms would move in to this uninhabited space?
  - What kind of organisms are the first to move in?
  - What factors affect the likelihood of different species migrating to the space?
- Predict what this ecosystem will be like in three months.
- Construct a hypothesis in relation to the possible development of an ecosystem here or related to a species you expect to move in early in the ecosystem development process.
- Mark out the space you intend to monitor.
- Record specific information about the dimensions of the area.
- Keep photographic and quantitative data (tabulated) about the organisms inhabiting the area. This includes population data using population monitoring techniques. GO TO ➤ Key knowledge page 155
- Check daily to monitor the space. Record observed changes. This will include quantitative data (tabulated) and qualitative data (descriptions).
- Daily monitoring can raise more questions. Record these. You do not have to answer all of the questions, but raising them demonstrates your awareness of the factors at play. They can also prompt adjustments to the procedure and the outcomes of the investigation, as well as launch further investigations.
- At the conclusion of the investigation you will prepare a scientific report detailing the investigation and its findings. Follow the protocols for a scientific report. The pages about Practical investigations in the Biology toolkit will be helpful. GO TO ➤ page ix

ISBN 978 1 4886 1931 1

MODULE 4 • REVIEW QUESTIONS

Mutiple choice

1 Organisms in ecosystems interact with one another and with their non-living environment in different ways. Interspecific competition is characterised by:

A competition between organisms of the same species for the same resources

B competition between organisms of different species for the same resources

C competition between organisms of the same species for different resources

D competition between organisms of different species for different resources

2 An organism's niche is:

A the place where it lives at a particular time

B its environment, including all of the factors that influence it during its lifetime

C its role in its habitat

D its relationship with individuals of its own and other species

3 The European carp (*Cyprinus carpio*) was introduced into Australian waterways around the mid-1800s. This invasive species has seriously impacted river systems, increasing erosion and water turbidity. It also outcompetes native fish species for resources, including food, leading to a significant decline in native fish populations. This kind of interaction between the introduced and native fish species is an example of:

A the competitive exclusion principle

B the competitive inclusion principle

C resource partitioning

D facultative symbiosis

4 Without any limiting factors, growth of a population will be exponential. Limiting factors can be density-dependent or density-independent. An example of a density-independent factor is:

A predation

B crowding

C natural disasters

D competition for resources

5 A stable ecosystem is self-regulating. While the populations of organisms fluctuate, various factors operate to keep them in check. The diagram below represents a self-regulating ecosystem.

abundant food supply

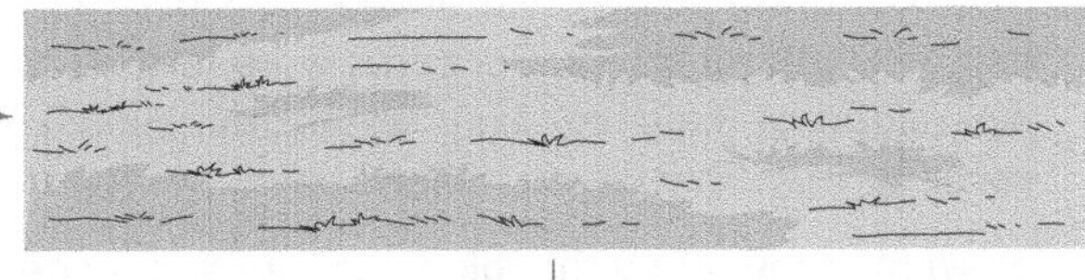

number of grazers increases through birth and immigration

grassland becomes overgrazed

decrease in food supply limits grazer numbers so they emigrate (leave the area) or die

The mechanism that maintains populations in such a way is called:

A interspecific competition

B intraspecific competition

C positive feedback loop

D negative feedback loop

6 Bioindicators provide important information about the relative health of ecosystems. Specifically, a bioindicator is:

A a biotic factor that indicates the presence of its own species in an ecosystem

B an organism that is sensitive to environmental changes

C a probe used to monitor biotic factors in ecosystems

D chemical indicators that provide information about factors that affect living organisms in ecosystems

Short answer

7 The diversity of species found in a waterway is a good indication of its health. When waterways become polluted, diversity declines as less tolerant species die out.

A study was conducted to identify the fish species present in a small stream near a proposed site for a paper manufacturing plant. Chlorine bleach is used to whiten paper during the manufacturing process. Chlorine bleach has a pH of around 12. Concerns were raised about how this might affect the fish in the waterway because the normal pH of the stream was approximately 6.5. An experiment was conducted to determine the pH-tolerance range of the fish living in the stream.

a What is meant by the tolerance range of a species?

b During the experiment 400 members of each species were raised in eight ponds. The first pond was maintained at a pH of 4.5, the second at 5.0, the third at 5.5, and so on. The last pond had a pH of 8.0. Fifty fish of each species were placed in each pond and their numbers monitored. After several weeks, the number of surviving fish in each pond was counted and conclusions about how they cope with various pH levels were drawn.

i Identify the independent variable in this experiment.

ii Identify the dependent variable in this experiment.

The results of the experiment are illustrated in the following table.

c Which species was shown to be most sensitive to changes in pH? Explain your answer.

		Fish species					
		Pike	Brown trout	Rainbow trout	Yellow perch	Bass	Salmon
pH	8.0						
	7.5						
	7.0						
	6.5						
	6.0						
	5.5						
	5.0						
	4.5						

d The fish population of each pond is easy to monitor as they are confined. However, monitoring fish in natural waterways is not so easy.

i Describe the challenges associated with monitoring fish populations in their natural habitat.

ii Outline a technique used to overcome these challenges when monitoring fish in natural waterways.

 ISBN 978 1 4886 1931 1

8 Geological evidence demonstrates that present-day Australia once belonged to the southern supercontinent of Gondwana made up of Australia, Africa, Madagascar, India, South America, New Zealand and Antarctica. One hundred million years ago the landmass was much further south with a much colder climate. Since then continental drift has been responsible for the break-up of the landmasses, with Australia beginning to break away from Antarctica around 85 million years ago and completing the separation approximately 30 million years ago. Since then the Australian continent has slowly moved northward to its present position. Its isolation as an island continent has driven the evolution of some of the most unique species of plants and animals in the world.

a Explain what is meant by the term biogeography.

b Describe changes in the abiotic features of the Australian landmass since its separation from the supercontinent of Gondwana.

c Outline the evidence that suggests a unique evolutionary pathway has occurred over the past 85 million years for the following organisms:

i sclerophyll plants

ii small mammals

9 Overexploitation of resources has pushed many species to extinction. Overexploitation occurs when resources are used at a rate higher than they can be replaced. This has been a problem in the commercial fishing trade, with overfishing practices seriously depleting fish populations. The consequences are far-reaching for the species in question, other organisms with which it interacts, and for the fishing industry that relies on these species for their livelihood.

Yellowfin tuna (*Thunnus albacares*) are an example of fish that have been overexploited by the fishing industry. They are migratory fish that follow the warmer waters of the Indian, Pacific and Atlantic oceans. They are highly sought after for commercial purposes. Yellowfin are predators of smaller fish, pelagic crabs and squid. They are preyed upon by sharks and marine mammals. Problematic fishing practices have included netting large schools of juvenile fish along with the target catch of larger, adult fish. Other marine life, including the vulnerable leatherback turtle, are also regularly caught in fishing nets.

a Draw a relationship web that illustrates the niche of the yellowfin tuna and the consequences of exploitative fishing practices.

b Explain the logic behind regulations that demand fish under a set threshold size must be returned to the water.

c Suggest two different strategies that could be implemented to address the yellowfin tuna problem. Explain the reasoning behind each approach you have described.

 ISBN 978 1 4886 1931 1

Extended response

10 Our growing human population impacts on local ecosystems, but also on the biomes and the biosphere. Our ever expanding need for living and agricultural space to support communities, and our growing consumption of energy to allow for transport, light and electronic goods such as computers and mobile phones, all put an increasing demand on the environment. The effects of this include deforestation, mining and industry. The consequences of these practices have included extinction of species, pollution and climate change, to name a few. In recent times it has become clear that in order for our global ecosystem to continue to sustain us, we must take steps to ensure its wellbeing.

The following table illustrates three animal species whose conservation status is currently listed as Critically Endangered, in large part due to human activity.

	Species			
		Common name	**Scientific name**	**Country of origin**
A		northern quoll	*Dasyurus hallucatus*	Australia
B		Amur leopard	*Panthera pardus orientalis*	southeast Russia northeast China
C		Sumatran orangutan	*Pongo abelii*	Sumatra

a Explain what is meant by 'Critically Endangered'.

b What is biodiversity conservation?

c Decide the most likely reason for the decline in numbers of each of these Critically Endangered species. Write the letter corresponding to the respective animals in the spaces provided below.

i	habitat loss due to harvesting of palm trees for palm oil	
ii	predation by feral cats and poisoning from preying on toxic cane toad	
iii	hunting for illegal wildlife trade	

d Discuss reasons for implementing conservation strategies that protect and maintain natural habitats and species from extinction. Include the following points of discussion in your response:

- the importance of biodiversity
- threats to biodiversity
- potential consequences of inaction
- the ecological value of ecosystems
- conservation strategies.

 ISBN 978 1 4886 1931 1